Control Applications of Microcomputers

Peter Mitchell

Edward Arnold

A division of Hodder & Stoughton

LONDON   BALTIMORE   MELBOURNE   AUCKLAND

© 1988 P. Mitchell

First published in Great Britain 1988

British Library Cataloguing in Publication Data

Mitchell, Peter
    Control applications of Microcomputers.
    1. Automatic control —— Data processing
    2. Microcomputers.
    629.8′312   TL223.M53

    ISBN 0–7131–3583–2

Typeset in Great Britain by TecSet Ltd, Wallington
Printed and bound in Great Britain, for Edward Arnold, the
educational, academic and medical publishing division of Hodder
and Stoughton Limited, 41 Bedford Square, London WC1B 3DQ
by J. W. Arrowsmith, Bristol.

Preface

To The Beginner

Let us suppose you were presented with several different computers and a small stepper motor, and you were given the problem of controlling this motor via one of the computers. You would no doubt have absolutely no idea where or how to start. You may ask yourself the following questions:

(a) What is a stepper motor and how does it work?
(b) Where do we connect this device to the computer?
(c) Does it matter which computer is used?
(d) Is it safe to connect the motor directly or could it damage the computer or vice versa?
(e) What sort of program is required for controlling the motor and are the instructions different from normal BASIC instructions?

Naturally I cannot answer these questions immediately here, but I will briefly discuss each to give you an outline of control using microcomputers. So to go over the questions again:

(a) What is a stepper motor? — It is simply a motor which steps in set amounts, like the second-hand on a clock. It may require say 48 steps to complete one revolution. This type of motor is discussed in detail in the text.
(b) Where is the device connected to the computer? — There are several plug sockets on microcomputers that appear to be unused. One may be for a printer, another for a floppy disc drive, but one is available for control purposes.
(c) Does it matter which computer is used? — No! The principles described in this book are relevant to all microcomputers. However, some are considerably better suited for control purposes than others, e.g. the 680X, and 650X as compared to the Z80 types.
(d) Is it safe to connect the motor directly? — The motor cannot be 'plugged' directly into the computer. The two are not compatible. All the information required to make them compatible is given in detail in this book. The point to realise is that computers can only handle very small currents and even the smallest of motors may require more current than the computer can deliver.
(e) What sort of program and instructions are required? — The programs in the book tend to be short (and simple) because in general only one device is being controlled at a time. Standard BASIC instructions are employed but use of a great many POKE and PEEK instructions is necessary. The BBC

Model B uses the indirection operator '?' instead of PEEK and POKE. Two numbers occur extremely often (PET, 59459 and 59471, BBC, 65122 and 65120 and Commodore 64, 56579 and 56577) in the programs and these are the addresses necessary to set the microcomputer to emit or receive signals, in control or monitoring applications. Again these are explained in detail later.

Two chapters in this book are devoted to assembly language for the 6502 microprocessor. One is a brief and simplified introduction, the other covers using this language for control purposes. The BBC B and PET both use the 6502 microprocessor and the Commodore 64 uses an up-dated version — the 6510. Therefore, the Commodore 64 also uses 6502 machine codes/assembly language.

When I became interested in control and interfacing techniques no one book seemed to answer all these questions and I struggled from book to book, magazine to magazine. The problems seemed large and endless and it took me a very long time to cover a small amount of theory. Having gone through this learning process the hard way I decided to try to answer the previous questions (and more) in this book. Although the subject is vast and complicated I have tried to make the introduction to this interesting field of technology as simple as possible. An assumption that you have no idea about control has been made, and the only pre-requisites are a genuine interest, a fair knowledge of BASIC and a bit of spare cash.

Acknowledgements

The author would like to express his thanks to his family for their patience throughout the preparation of this book. Grateful thanks are also due to Julie and Linda Witowski for their much appreciated assistance.

Peter Mitchell
1987

Contents

1

The microprocessor

Integrated circuits

The field of electronics has changed dramatically over the past 50 years. Initially it was dominated by electronic valves, such as the triode, but these were soon superseded by smaller, cheaper semiconductor devices like the transistor. Nowadays integrated circuits, or ICs as they are popularly known, are used in great numbers although transistors and other discrete components are still in great demand.

Integrated circuits are tiny circuits manufactured on the surface of a small piece of pure silicon. The finished circuit is then located in a plastic package with connecting pins. These circuits are classified according to the number of components on the silicon chip. The classification is as follows:

(a) SSI (small scale integration) — up to 10 components on a chip (see Figure 1.1).
(b) MSI (medium scale integration) — more complex circuits with up to 100 components.
(c) LSI (large scale integration) — these have up to 1000 components and are very complex circuits.
(d) VLSI (very large scale integration) — tens of thousands of components on a single chip. The microprocessor, and many ICs used along with it, fits into this category

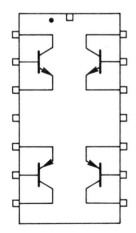

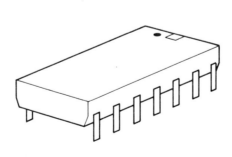

Fig. 1.1 An SSI circuit and its package

Hundreds of different ICs are manufactured and most have specific or dedicated function. The microprocessor however, is not a dedicated device as it can be programmed to function to suit particular applications.

The microprocessor

More and more control and monitoring applications are being taken over by the microprocessor. The reasons for this are simple. Microprocessors are small, reliable and cheap, but above all they are programmable. It is because they are programmable that they can be used as universal controllers rather than, as stated already, dedicated controllers for specific applications. The same microprocessor could be used in an automatic washing machine, a CNC lathe or a computer.

Figure 1.2 is the 6502 microprocessor used in the BBC B and PET micro-computers. It has 40 pins connected to the processor chip.

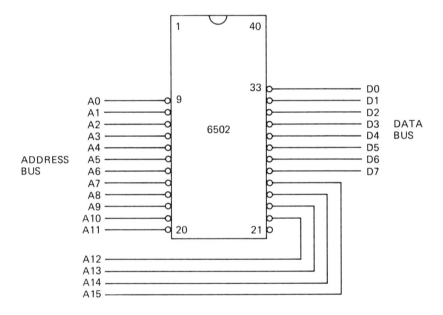

Fig. 1.2 6502 microprocessor

The microprocessor is a **digital** device and in digital electronics there are only two defined states. The states are given the values of 0 (representing no signal) and 1 (representing a signal or a voltage). So on each of the 40 pins there will be either a '0' or a '1'.

Buses

A bus is a set of parallel conductors used for carrying signals that are related. Figure 1.2 shows an address bus with sixteen lines or conductors and a data bus with eight lines. There is also a control bus which deals with the internal microelectronic systems of the computer. It is not particularly relevant to the control of external devices (at this level) and so is not discussed in this text.

Address bus

In order for the microcomputer system to know where data is stored in memory, each memory location must be allocated its own unique identification label. These labels are called addresses. If the user of the microcomputer wishes to read the data stored in a particular memory location he can do so by specifying the address in memory. The micro-system then places the necessary signals on the address bus and retrieves the data from memory.

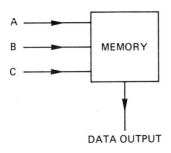

Fig. 1.3 Simple memory system

Figure 1.3 depicts a simplified memory system with a 3 line address bus. On each address line there can be either a '1' or a '0' and the possible signal combinations on the address bus are shown in Table 1.1.

Table 1.1

A	B	C	Memory Address
0	0	0	0
0	0	1	1
0	1	0	2
0	1	1	3
1	0	0	4
1	0	1	5
1	1	0	6
1	1	1	7

The table shows that with such an address bus it is possible to access 8, ie, 0 to 7, different memory locations. We can calculate the number of addresses that can be accessed with a particular 'size' of bus as follows:

$$2^{3 \text{ address lines}} = 8 \text{ addresses}$$

A 16 line address bus can access 2^{16} (ie. 65536) different addresses in memory. Such a system is limited to addressing this amount of memory and is referred to as 64K. This assumes of course that the microcomputer has a memory of such size to begin with.

Data bus

The 8 line data bus carries data words made up of 1s and 0s. Each 1 and 0 is called a bit and eight bits make a word. Such a data bus is capable of carrying 2^8 (or 256) different data words. Figure 1.4 shows the data bus and a typical data word.

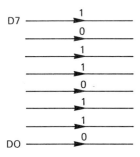

Fig. 1.4 The data bus carrying a data word

Data buses must carry data words to and from the memory and input/output sections and this means that the bus must be bi-directional.

2

Number systems

Three very common number systems are used in computing, and these are denary (decimal), binary and hexadecimal. Each of these has its own areas of importance.

Denary   Everyone uses this system in everyday life and naturally it is used in computing. Its base is 10 and it uses the numbers 0 to 9.

Binary   Whether you like it or not a sound knowledge of binary is essential for anyone involved in computer control. Its base is 2 and it uses the numbers 0 and 1 only. The 1 and 0 states are very useful because they can be used to represent a signal (or a voltage) or no signal (or no voltage) respectively.

Hexadecimal   This system is used because it is a very short and convenient way of representing binary numbers. For example the number 111101011111 in binary is F5F in hexadecimal. Hexadecimal has a base of 16 and uses numbers from 0 to 15.

System identification

In order that the microcomputer can distinguish each number system prefixes are used. A typical set of prefixes *could* be:

	Prefix
Denary	non
Binary	%
Hexadecimal	$

These would be used as follows:

255	Denary
%11111111	Binary
$FF	Hexadecimal

Note: these prefixes are not standard and will vary from machine to machine — for example the prefix used on the BBC micro to indicate hexadecimal numbers is &.

Denary (decimal) systems

The number 5638 means $5 \times 1000 + 6 \times 100 + 3 \times 10 + 8 \times 1$ and can be tabulated as follows:

1000	100	10	1
5	6	3	8

The table is set in columns — each ten times greater than the previous. Starting from the left hand column $1000 = 10^3$, $100 = 10^2$, $10 = 10^1$ and so $1 = 10^0$. Retabulating this we have

1000 10^3	100 10^2	10 10^1	1 10^0
5	6	3	8

Note the peculiar right hand column — $1 = 10^0$. This is from the rule that anything to the power of 0 is one. So for example:

$$6^0 = 1$$
$$96^0 = 1$$
$$E^0 = 1$$

This can be proved fairly easily; take the example

$$\frac{25}{25} = 1$$

This can be rewritten as $1 = \dfrac{25}{25} = \dfrac{5 \times 5}{5 \times 5} = \dfrac{5^2}{5^2} = 5^{2-2} = 5^0$

Binary system

Binary numbers can be broken down in much the same way as denary numbers. For example the binary number 11001 represents, reading from right to left, a one, no 2s, no 4s, one 8 and one 16. This gives a total of 25 in denary. The table below illustrates the relationship between binary and denary.

	2^5 (32)	2^4 (16)	2^3 (8)	2^2 (4)	2^1 (2)	2^0 (1)
Denary						
Binary		1	1	0	0	1

Hexadecimal system

Hexadecimal numbers are 0 to 15. However, the numbers 10 to 15 are actually represented by the letters A to F. A complete list of basic hexadecimal numbers and their denary equivalents are shown below:

Hexadecimal	Denary
0	0
1	1
2	2
3	3
4	4
5	5
6	6
7	7
8	8
9	9
A	10
B	11
C	12
D	13
E	14
F	15

An hexidecimal (hex for short) number such as 2CA can be broken down as follows:

16^3 4096	16^2 256	16^1 16	16^0 1
	2	C	A

This means $2 \times 256 + C(12) \times 16 + A(10) \times 1 = 714$. Therefore, 2CA hex = 714 denary.

Conversion of numbers from one system to another

(a) Binary to Denary

Example
As was previously stated binary numbers are made up of digits called bits and each bit has its own denary weighting. Starting from the right hand end of a binary number the first bit is weighted as 1, the next 2, the third 4 and so on. To

convert binary to denary requires writing the denary weightings above the binary bits (as shown below) and then adding all the denary numbers where the binary bit is a '1'.

```
denary   16   8   4   2   1
binary    1   1   0   1   1
```

The number above has one 16, one 8, no 4, one 2 and one 1.

```
    16
     8
     2
     1
   ____
    27
```

Therefore, binary 11011 = denary 27.

Exercise 2.1
Convert the following binary numbers to denary:

(a) 10001000 (b) 100101 (c) 01111000000

(b) Denary to Binary

This involves dividing the decimal number by 2 and recording the remainder, and repeating this again and again until zero is reached. The example below illustrates the principle.

Example
Convert denary 176 to binary.

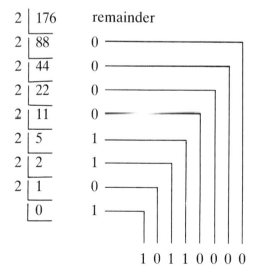

```
2 | 176      remainder
2 | 88        0
2 | 44        0
2 | 22        0
2 | 11        0
2 | 5         1
2 | 2         1
2 | 1         0
  | 0         1
```

1 0 1 1 0 0 0 0

Therefore, denary 176 = binary 10110000.

Exercise 2.2

Convert the following denary numbers to binary:

(a) 59 (b) 211 (c) 315

(c) Binary to Hexadecimal

This is suprisingly easy to do because all that is required is to group the binary bits into groups of four starting from the right hand end. Each group is then converted into the equivalent hex number.

Example
Convert binary 11101011100111 into hexadecimal.

Grouped into fours:

11	1010	1110	0111
3	A	E	7

Therefore, binary 11101011100111 = hex 3AE7.

Exercise 2.3
Convert the following binary numbers to hexadecimal:

(a) 11110011 (b) 110111 (c) 111111011000

(d) Hexadecimal to Binary

This conversion is simply a reversal of the previous one. Each hexadecimal digit is converted into its equivalent (four) binary bits.

Exercise 2.4
Convert hex 3F92 into binary.

3	F	9	2
0011	1111	1001	0010

Therefore, hex 3F92 = binary 0011111110010010.

Example
Convert the following hex numbers to binary:

(a) 31 (b) A3 (c) 4E8

(e) Decimal to Hexadecimal

This is a very similar process to converting decimal to binary, except the decimal number is divided by 16 rather than by 2. Again the remainder column provides the solution.

Example
Convert denary 4056 to hex.

```
                    Remainder
   16 | 4056
   16 | 253        8 (8)
   16 | 15         13 (D)
    0  0           15 (F)
```

Therefore, denary 4056 = hex FD8.

Exercise 2.5
Convert the following denary numbers to hex:

(a) 116 (b) 233 (c) 55

(f) Hexadecimal to Denary

The simplest way to explain this process is by example. Suppose we wish to convert hexadecimal number FF to denary. A table can be drawn up as follows:

16^2 (256)	16^1 (16)	16^0 (1)	Denary
	F	F	Hex

This means there are F (or 15) × 1 and F (or 15) × 16

$$F(15) \times \ 1 = 15$$
$$F(15) \times 16 = 240$$

Therefore FF = 255

Example
Convert 7C2D hex to denary.

16^3 (4096)	16^2 (256)	16^1 (16)	16^0 (1)
7	2	C	D

This gives

$$
\begin{array}{lrcr}
\text{D}(13) \times & 1 & = & 13 \\
2 \times & 16 & = & 32 \\
\text{C}(12) \times & 256 & = & 3072 \\
7 \times & 4096 & = & 28672 \\
\hline
& & = & 31789
\end{array}
$$

Exercise 2.6
Convert the following hex numbers to denary:

(a) 60 (b) 276 (c) 4E8

3

General interfacing techniques

The circuits described in detail in this book are suitable for the BBC Model B, Commodore 64 and PET microcomputers. However, the theory is relevant to all machines and the circuits can be fully utilised by other micros by slight modification of the adaptor circuit discussed. Alteration to this circuit involves rewiring it to comply with the microcomputers user port arrangement.

The user port

If we are to control devices, then the first thing we need to know is where do we connect them to the microcomputer. The answer is — the user port.

The user port is a plug socket whose location is generally at the base of the computer. The port positions for the BBC Model B, Commodore 64 and PET are illustrated in Fig. 3.1.

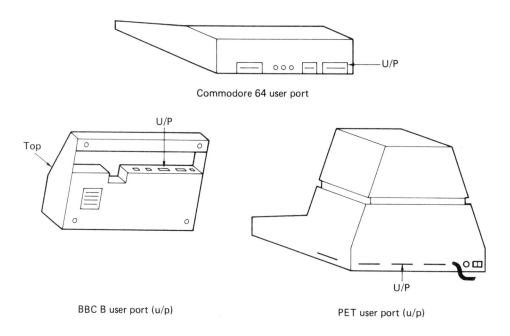

Fig. 3.1 User port location

Figure 3.2 is a closer look at the user ports for these three micros and shows the pin configurations.

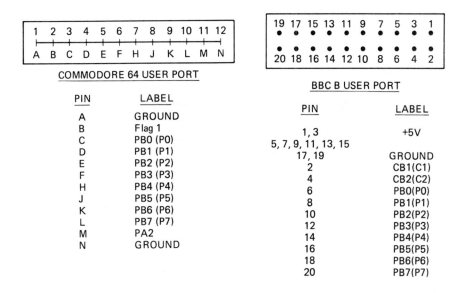

COMMODORE 64 USER PORT

PIN	LABEL
A	GROUND
B	Flag 1
C	PB0 (P0)
D	PB1 (P1)
E	PB2 (P2)
F	PB3 (P3)
H	PB4 (P4)
J	PB5 (P5)
K	PB6 (P6)
L	PB7 (P7)
M	PA2
N	GROUND

BBC B USER PORT

PIN	LABEL
1, 3	+5V
5, 7, 9, 11, 13, 15 17, 19	GROUND
2	CB1(C1)
4	CB2(C2)
6	PB0(P0)
8	PB1(P1)
10	PB2(P2)
12	PB3(P3)
14	PB4(P4)
16	PB5(P5)
18	PB6(P6)
20	PB7(P7)

PET USER PORT

PIN	LABEL
A	GROUND (0V)
B	CA1(C1)
C	PA0(P0)
D	PA1(P1)
E	PA2(P2)
F	PA3(P3)
H	PA4(P4)
J	PA5(P5)
K	PA6(P6)
L	PA7(P7)
M	CB2(C2)
N	GROUND (0V)

Fig. 3.2 Pin identification

User ports can be programmed to emit or receive signals. These signals to or from the port are at TTL(Transistor Transistor Logic) levels.

TTL levels:
logic 0 = no signal ≃ 0 volt
logic 1 = a signal ≃ 5 volt

The three micros discussed have eight programmable input/output lines (P0–P7). It is therefore possible to control or monitor up to eight separate devices. There is naturally a ground (or 0V) line and there are two other lines, C1 and C2, which have a special function and are used for a technique known as 'handshaking' which is described later.

Connectors/plugs

Connectors (or plugs) to fit the user port are readily available generally, and are fairly cheap. A typical design of connector for the Commodore 64 is illustrated in Fig. 3.3. Naturally great care must be taken when soldering cables to the connectors to ensure correct connections to the pins (in accordance with the pin configuration). The top of the plug should also be marked.

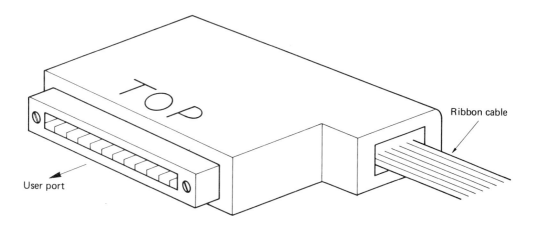

Fig. 3.3 Typical Commodore connector

Note that the ribbon cable should have at least eleven lines: P0/P7, 0V, CB1, CB2 for the BBC B. Similarly for the Commodore 64 and PET.

Adaptors

In order to make the process of wiring different devices to the user port convenient, an adaptor can be made. This is simply a plug and a box incorporating 4mm sockets. Wiring will be simple, flexible and fast, and no soldering is required when devices are changed. The added advantage is that only one adaptor and user port connector will be required however many devices you have at your disposal (the term devices is used to include all items to be controlled or monitored).

The cost is low and the general layout and list of equipment for the adaptor is given in Fig. 3.4. The procedure is to solder the ribbon cable to the user port connector in accordance with the user port configuration; 4mm sockets can then be set in the top of the box and the ribbon cable connected to these sockets.

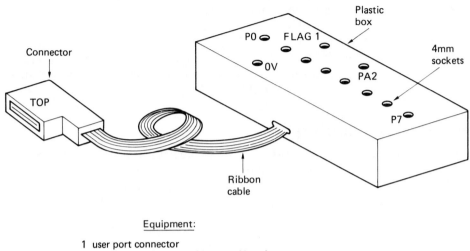

Equipment:

1 user port connector
1 plastic box (150mm × 80mm × 40mm)
1 piece ribbon cable (1m long)
11 4mm sockets.

Fig. 3.4 Commodore adaptor

One precaution is to ensure that some mechanical fastening of the cable inside the box is arranged otherwise the cable may work loose with continual use.

Interfacing

The interface

To control (or monitor) devices using the computer we require a program (software) and a suitable connection between the computer and the device to be controlled. This 'suitable connection' is electronic circuitry which makes the computer and the device compatible, and is known as the interface. Figure 3.5 shows a block diagram of the required system.

Fig. 3.5 Block diagram of a control/monitor system

Why is interfacing necessary?

In many cases the power available directly from the computer is not enough to drive the devices to be controlled. A robot arm for example, requires a much greater current than the computer can deliver and so an individual power supply for the arm has to be used and switched into action by the computer.

Signals also enter the computer from devices. If for example we wished to use the computer to assist in the measurement of weight, or voltage or temperature etc., then the incoming signals *may* be too large or too small. Again circuitry is required to bring the signals to a suitable level.

Transistor switching

Consider the problem of controlling a small electric motor and, to make it simpler, let us assume that we require only to switch the motor on and off. This particular situation arises in many applications, for example, in automatic washing machines.

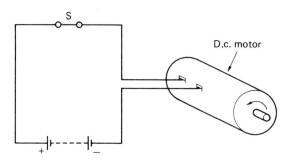

Fig. 3.6 Basic d.c. motor circuit

Figure 3.6 shows the normal circuit required to switch a small d.c. motor on and off. Obviously to control this motor using a computer, it is only necessary to use one of the eight control lines and the 0V line (see Fig. 3.7). Any of the eight lines P0–P7 could be used but P0 has been selected. An output signal switches the motor on and no signal means the motor is off. Note that current flows along P0 and back along the 0V line.

The circuit (Fig. 3.7), however, is unsuitable for two reasons:

(a) The user port is not capable of delivering the current required to drive the motor. A small motor may require say 500 mA ($\frac{1}{2}$ A) but the user port capacity is well below this.

(b) The motor with its low internal resistance would try to draw the current it requires, thereby *possibly* damaging the user port. This is similar to shorting a screwdriver across the terminals of a car battery.

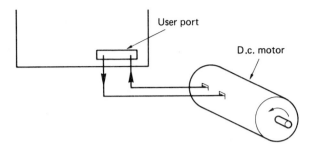

Fig. 3.7 D.c. motor control by comsputer — circuit unsuitable

It is necessary, then, to provide the power required by the motor (or whatever item) and simultaneously to protect the user port. A switching mechanism must be employed so that the small output current switches on a larger current to the motor from an external power source. Several components are available to do this, e.g. transistors and relays.

Transistors are three-legged components, and a typical one with its circuit diagram symbol is shown in Fig. 3.8.

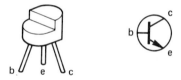

Fig. 3.8 A typical transistor and its symbol

To understand the principle of transistor switching it is not necessary to study the complex mechanics of transistor operation. All that is required is to know that the three legs are connected to the transistor's base(b), emitter(e) and collector(c) respectively, and a small voltage (or current) applied to the transistor's base increases conduction (reduces resistance) between the collector and emitter. Figure 3.9 explains this in a little more detail. Although there is a complete circuit ABCD, the lamp in Fig. 3.9(i) will not light because the transistor is switched off. A small current i applied to the base (Fig. 3.9(ii)) opens the flood gates (switches the transistor on) and current can then flow from the battery, around the circuit ABCD.

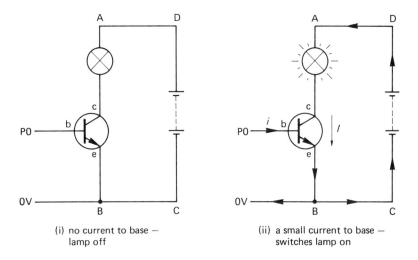

(i) no current to base —
lamp off

(ii) a small current to base —
switches lamp on

Fig. 3.9 Transistor as a switch (i) no current to base — lamp off, (ii) a small current to base switches lamp on

Transistor switching is ideally suited for control applications because only a minute current (generally microamps – millionths of an amp) is necessary to switch on much larger currents.

Let us now return to the problem of controlling the motor. The lamp in the circuit, Fig. 3.9, could be replaced by the motor but it is quite likely that a single transistor has not the capacity to cope and so a second transistor is placed in series. The final arrangement connected to the user port is given in Fig. 3.10. A small current (µA) from the user port is applied to the base of the first transistor TR1 and turns it on. A larger current (mA) flows from the battery and down through R1 and TR1 and is directed to the base of a more powerful transistor TR2. An even larger current from the battery can now flow down through the motor (turning it on) and through TR2.

The diode D1 is connected across the motor to protect the circuit against induced (or back) e.m.f.s produced when the motor is switched. The magnitude of the e.m.f.s (voltages) produced can be such that other parts of the circuitry could be damaged. Introduction of the diode means that these e.m.f.s are controlled by directing them around the circuit 1, 2, 3 and 4 (see Fig. 3.10).

The concept of transistor switching is relatively simple but selection of the correct transistor and accompanying resistors for particular applications is not so simple.

Output and input circuits

To remove the problem of interfacing each item to be controlled or monitored, a general purpose output circuit and a general purpose input circuit can be produced.

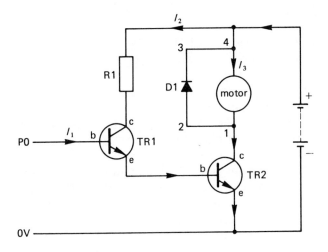

Fig. 3.10 Motor control using transistors

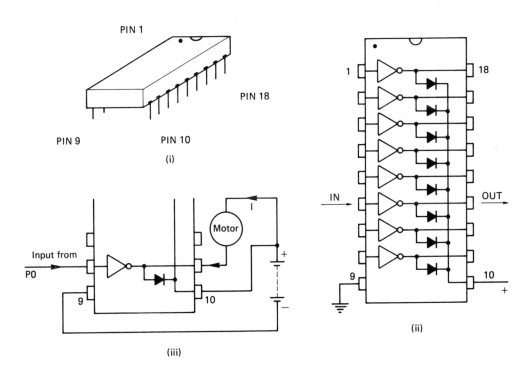

Fig. 3.11 Darlington driver, (i) integrated circuit, (ii) pin configuration, (iii) illustration of one section in use. Note that each of the darlington sections contains a diode which provides protection against back e.m.f.s.

Output circuit

This circuit should be capable of driving most, if not all, the devices you wish to run from the user port. Its operation is similar to that in Fig. 3.10 — a double transistor arrangement known as a '**darlington driver**' is utilised. Sixteen transistors are required for the eight output lines but fortunately all these have been incorporated on a single integrated circuit (a chip) as shown in Fig. 3.11.

A complete circuit diagram is given in Fig. 3.12 along with a list of the components required. I advocate the use of a plastic box with 4mm sockets for each external connection for ease of use. The LED and its accompanying resistor are optional as this simply indicates power on and off. A maximum of 50V and 0.5A *can* be handled by this driver, but if greater currents and voltages are to be controlled, relays can be introduced into the circuit. To keep the circuit as compact as possible if relays are to be used reed relays could be utilised (Fig. 3.13). The coil of the relay would be connected between A and B on the output circuit. A section of the modified output circuit is shown in Fig. 3.14.

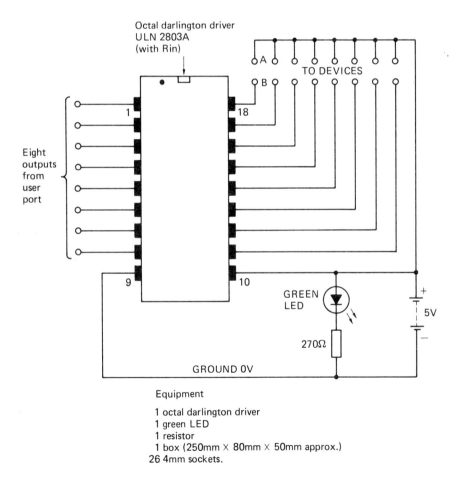

Equipment

1 octal darlington driver
1 green LED
1 resistor
1 box (250mm × 80mm × 50mm approx.)
26 4mm sockets.

Fig. 3.12 Output circuit

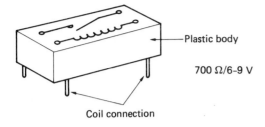

Fig. 3.13 Typical reed delay

Plastic body

700 Ω/6–9 V

Coil connection

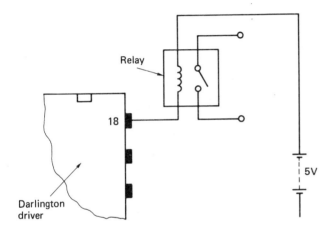

Fig. 3.14 Modified output circuit

Relay

18

5V

Darlington
driver

   Whichever option is chosen, the number of components to be used is very
small indeed and so the reliability is high. If the output circuit incorporating
relays is to be used then the suggested reed relays are suitable. If, however, you
wish to use more powerful relays, it is important to consider the coil resistance
and the current the coil will draw.

Input circuit

Input signals to the user port must be in the TTL range. Logic 0 is represented by
approximately 0V and logic 1 by voltages greater than about 3.25V. To protect
the user port from excessive input voltages, an **opto-isolator chip** can be used.

Opto-isolators

Opto-isolators, sometimes called photo-couplers, allow signals to pass from one
system to another although there is no physical connection inside the device.
They consist of an LED (light emitting diode) and a photo transistor (see Fig.
3.15). LEDs are described in detail in Chapter 4. An incoming signal passes

through the LED which illuminates. The emitted light hits the base of the photo transistor and switches it on. Current can then flow through the transistor to the user port. Thus the user port is totally isolated from the original incoming signal.

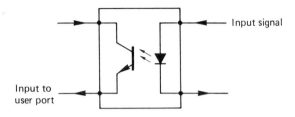

Fig. 3.15 An opto-isolator

None of the circuits in this book require the use of a general input circuit, but it is possible you may require such a device. Figure 3.16 is a diagram of a circuit to provide one input. To produce a circuit which can handle more inputs simply requires an increase in the number of these (Fig. 3.17). Both IC1 and IC2 can accommodate four inputs, so only two each of these ICs would be required to handle eight inputs.

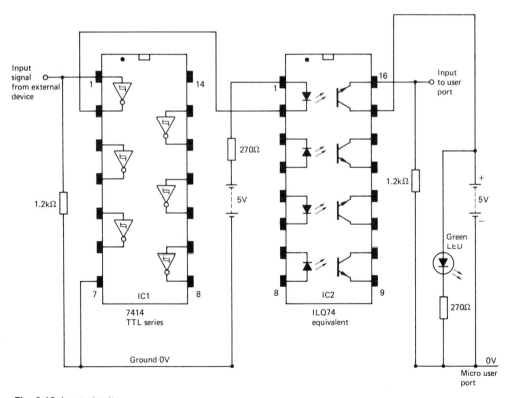

Fig. 3.16 Input circuit

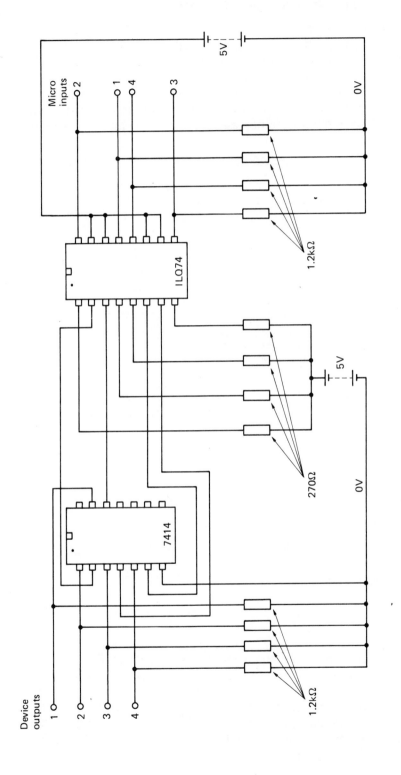

Fig. 3.17 Complete input circuit

General use of input/output circuits

Although the input and output circuits are completely separate, there is no reason why they cannot be used simultaneously. In fact, in many cases, this will be necessary. For example, several devices can be controlled whilst simultaneously other conditions are being monitored. A block diagram illustrating how the circuits would be used is shown in Fig. 3.18. It should be remembered that the 0V (ground) from every device, user port and control board (input/output) are all connected together and a 5V power supply is required by the input and output circuits.

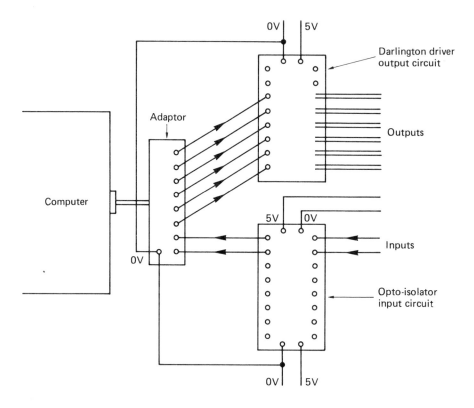

Fig. 3.18 Use of the input and output circuits

On some microcomputers, such as the BBC Model B, there is a 5V line available at the user port; this could, in some cases, be used instead of the external power supplies, but again the current from this line is very limited and I suggest you leave this line alone completely. It is advisable also, until you become an expert in the microelectronics field, to make full use of the input circuit when handling incoming signals so as to provide complete isolation and protection from the experimenters' inevitable mistakes.

Miscellaneous equipment

It has been stressed several times that to make wiring simple, 4mm sockets in plastic boxes should be used. If you have decided to follow this advice, then you will require numerous (20 or more) leads of different lengths, with 4mm plugs at each end. The most useful types are those which allow stacking as shown in Fig. 3.19.

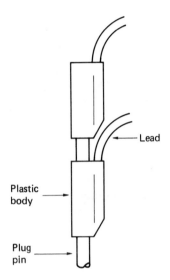

Lead

Plastic body

Plug pin

Fig. 3.19 4 mm plugs

Multimeters — meters to measure voltage, current and resistance — are essential to the builder of electronic circuits and satisfactory types are relatively inexpensive.

4

Output signals

The signals which are emitted from the user port are 5V (strictly > 3 volts approx.). Although very little power can be drawn from the user port, a simple circuit to investigate user port output signals can be made quite easily. This circuit involves using eight LEDs.

LEDs (Light Emitting Diodes)

An LED is an electronic component (see Fig. 4.1) which illuminates when a small current, about 20mA, passes through it. Their cost is very low and they can be purchased in several colours — red, yellow, green.

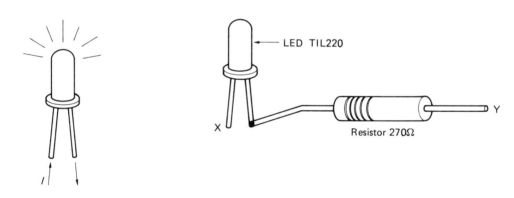

Fig. 4.1 A typical LED   **Fig. 4.2** LED test circuit

A simple circuit to observe the action of an LED can be made in just a few minutes. The components required are shown in Fig. 4.2.

All that is required now is to connect a small battery, say $4\frac{1}{2}$V, to X and Y. Connection of this battery one way illuminates the LED, whereas reverse connection does not. This means that the diode allows current to flow in one direction only. It is important, therefore, when wiring the following circuits that the LEDs are connected the right way round.

The circuit symbol for an LED is given in Fig. 4.3 and current must flow downwards in the diagram representation (anode 'a' to cathode 'k') to illuminate the LED. On an actual LED the cathode can be identified by a 'flat' on the body.

Fig. 4.3 Circuit symbol for an LED

LED control circuit

Controlling LEDs from the user port is by far the most useful, satisfactory and efficient way of learning about control. Figure 4.4 is the complete LED control circuit. Its operation utilises the switching of transistors described in Chapter 3. LED illumination should be satisfactory but because all the components are operating well within their capacities poor illumination can be overcome by a slight increase in battery voltage. The cost of the circuit is fairly low; similar ready made circuits can be purchased but naturally their cost is much greater.

To build the circuit, little practical experience is required but a few precautions are:

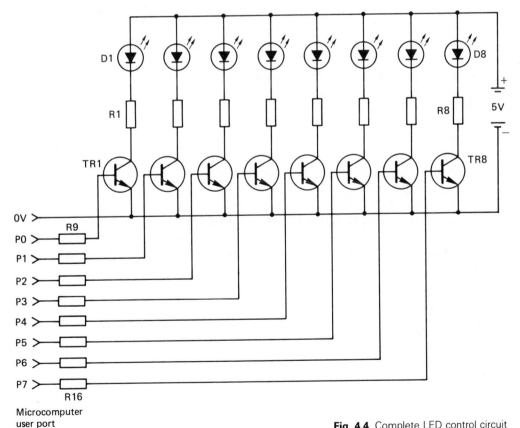

Fig. 4.4 Complete LED control circuit

(a) Ensure, as already has been mentioned, that the LEDs are connected the correct way.
(b) Ensure the transistors are connected correctly (see Fig. 4.5)
(c) Do not overheat the transistors when soldering them, as damage can result.

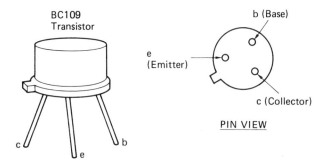

Fig. 4.5 BC109 transistor lead identification

Before starting production, however, I advise you to refer to Fig. 5.9 for a modified circuit. This circuit, although more complicated, can be used for output and input investigation.

A list of equipment required to build the circuit in Fig. 4.4 is as follows:

Resistors R1–R8   330Ω $\frac{1}{4}$ watt
        R9–R16   1kΩ

LEDs   D1–D8   RED TIL220
Transistors TR1–TR8   BC109

Printed circuit board approx. 120mm × 80mm
Tinned copper wire
4mm plugs (11 off)
4 small screws and locknuts to provide legs for the circuit.

The corner of the circuit board can be drilled and legs fitted as indicated in Fig. 4.6.

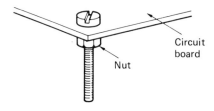

Fig. 4.6 Leg arrangement for circuit board

Connection of this circuit to the user port is via the adaptor described in Chapter 3, and the completely connected control circuit is given in Fig. 4.7. Note that *all* the 0V lines are connected together and this includes the negative battery terminal.

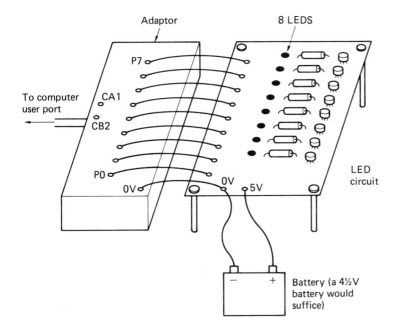

Fig. 4.7 LED circuit connected to microcomputer

Once the circuit (Fig. 4.7) is connected to the user port and the computer is switched on programming can commence. Enter and run the following program:

PROG. 4.1

BBC	Commodore 64	PET
10 ?65122 = 255	10 POKE 56579,255	10 POKE 59459,255
20 ?65120 = 15	20 POKE 56577,15	20 POKE 59471,15
30 END	30 END	30 END

On running the program it will be observed that four LEDs are on and four are off. Excluding the END statement the LEDs have been 'controlled' with a two line program. Let us examine both lines in detail.

Data Direction Register (DDR)

Line 10 in the program above tells the computer which control lines will be output and which will be input — it sets the **direction of data**.

The Data Direction Register (**DDR**) is an 8 bit register in memory (Fig. 4.8). If all the bits in this register are set at '1' then all the control lines will be outputs. If all the bits are set at '0' then all the lines will be inputs. Any combination of inputs and outputs can be programmed.

Fig. 4.8 The Data Direction Register

In our program line 10 (for the BBC) read ?65122 = 255. 65122 is the DDR. ?(POKE) and 255 tells the computer to place 255 decimal (11111111 binary) in the DDR. This means that all the bits in the DDR are set as '1' and therefore all the control lines are set as outputs. Line 10 for the Commodore 64 program is very similar, the only difference being the address of the DDR. Although the BBC line 10 appears to be different it achieves exactly the same result. Figure 4.8 represents the DDR for our program.

It must be appreciated that *no* output is provided by the computer at this stage, we have only set the **direction** that data will flow.

If line 10 was changed to POKE 59459,15 then only the first four control lines (P0–P3) would be outputs, and P4–P7 would be inputs.

Data Register (DR)

This is sometimes called the **input/output register**.

Once the Data Direction Register has been set, (and it is best to form the habit of setting it at the start of the program) a desired output can be sent.

Line 20 in Program 4.1 was ?65120 = 15 for the BBC B. 65120 is the **data register**. To put it crudely the bits in this register are the signals entering or leaving the computer.

?(POKE) and 15 means place 15 decimal (00001111 binary) in the DR. The number is converted into its equivalent binary bits, and output signals are produced from the user port. Again line 20 for the Commodore 64 and PET programs produce the same result.

Rerun the program and change the number from 15 to any number between 0 and 255. Then try numbers outside this range.

This program is rather poor, because the number in line 20 has to be changed each time. The following program makes use of an INPUT statement and is much better to use.

PROG. 4.2

BBC	Commodore 64	PET
10 ?65122 = 255	10 POKE 56579,255	10 POKE 59459,255
20 INPUT A	20 INPUT A	20 INPUT A
30 ?65120 = A	30 POKE 56577,A	30 POKE 59471,A
40 GOTO 20	40 GOTO 20	40 GOTO 20

Line 10 sets all the lines as outputs.
Line 20 allows the programmer to input any desired value.
Line 30 produces the output to the LEDs.
Line 40 starts the program again.
   Try the following numbers for A: 1, 2, 4, 8, 16, 32, 64, 128. Each of these numbers lights an individual LED. Why?

Time delays

The next step is to use a time delay to switch the LEDs off after a period of time. The flowchart for our next program is illustrated in Fig. 4.9.

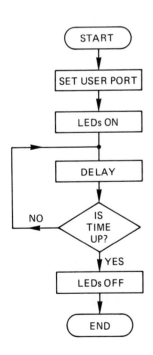

Fig. 4.9

A time delay could consist of asking the computer to do umpteen calculations; unfortunately the number of calculations would be large and time delays would be fairly unpredictable. The most organised and controlled way is to use a counting loop.

It takes the PET approximately 100 seconds to count to 65000.* Using this as a yardstick, we can predict the number required in the counting loop to provide a suitable delay. A loop such as:

    FOR I = 0 to 4000
    NEXT I

is an ideal sort of BASIC delay.

The program below illuminates the LEDs for a period of time and then switches them off.

PROG. 4.3 Use of time delay

BBC	Commodore 64	PET
10 ?65122 = 255	10 POKE 56579,255	10 POKE 59459,255
20 INPUT A	20 INPUT A	20 INPUT A
30 ?65120 = A	30 POKE 56577,A	30 POKE 59471,A
40 FOR I = 0 TO 4000	40 FOR I = 0 TO 4000	40 FOR I = 0 TO 4000
50 NEXT I	50 NEXT I	50 NEXT I
60 ?65120 = 0	60 POKE 56577,0	60 POKE 59471,0
70 GOTO 20	70 GOTO 20	70 GOTO 20

Line 10 sets all lines as outputs.
Line 20 allows the input of a desired value.
Line 30 switches selected LEDs on.
Lines 40/50 provide a time delay.
Line 60 switches all LEDs off.
Line 70 starts the program again.

Although I advocate the use of REM statements, I have omitted these to keep the program as short as possible.

Flashing LEDs

A progression from program 4.3 is to produce a program to make the LEDs flash on and off continuously. Here two delays will be required; one while the LEDs are on and one while they are off. If the times for both are equal then a single delay routine can be housed in a subroutine as illustrated in program 4.4. Often in a control cycle a delay will be used several times and naturally using a subroutine to hold the delay saves time for the programmer.

* It takes the PET so long because its BASIC interpreter is so slow. Machine code would take a fraction of a second to count to 65000.

PROG. 4.4 Flashing LEDs

BBC	Commodore 64	PET
10 ?65122 = 255	10 POKE 56579,255	10 POKE 59459,255
20 INPUT A	20 INPUT A	20 INPUT A
30 ?65120 = A	30 POKE 56577,A	30 POKE 59471,A
40 GOSUB 500	40 GOSUB 500	40 GOSUB 500
50 ?65120 = 0	50 POKE 56577,0	50 POKE 59471,0
60 GOSUB 500	60 GOSUB 500	60 GOSUB 500
70 GOTO 30	70 GOTO 30	70 GOTO 30
500 REM DELAY	500 REM DELAY	500 REM DELAY
510 FOR I = 0 TO 500	510 FOR I = 0 TO 500	510 FOR I = 0 TO 500
520 NEXT I	520 NEXT I	520 NEXT I
530 RETURN	530 RETURN	530 RETURN

Line 10   Sets all lines as outputs.
Line 20   Allows any number to be input to select LEDs on.
Line 30   Switches on selected LEDs.
Line 40   Jumps to a delay routine.
Line 50   Switches all LEDs off.
Line 60   Jumps to delay routine.
Line 70   Starts cycle again.

Line 500 Start of delay.
Line 510 ⎫
Line 520 ⎭ Counting loop.
Line 530 Return to original main program

The LEDs are now under the programmer's control; all that is required is a little more practice.

Exercise programs (solutions on p. 161)

Write the following programs for the LED circuit.
4.5 to count in binary form 0 to 255
4.6 to light each LED one after the other — sequentially
4.7 to light LEDs in pairs, one pair after another.

5

Input signals

Microprocessors are being used more and more for the control and monitoring of everyday items. Two areas where the use of microprocessors is spreading are motor cars and washing machines. The main reason is that the microprocessor, unlike conventional controllers, can be used in any control application because it can be programmed to suit the system's requirements. It can control fans, heaters, motors and even complete machines such as the modern CNC lathes. (Note: CNC — Computerised numerical control.)

In many systems a control sequence may be triggered by some incoming signal. For example, the heating element of some system could be switched on and off by the microprocessor — but when should the heater be off or on? The answer is of course when the microprocessor receives a signal to indicate that the temperature has reached a certain level. These signals must be produced by a component sensitive to the quantity (such as temperature variation) being monitored. Sensors and transducers are the terms used to describe these components. Several common sensors/transducers are discussed below.

Transducers

Loudspeakers and microphones are very common transducers (Fig. 5.1). One converts electrical signals into sound and the other converts sound into electrical signals.

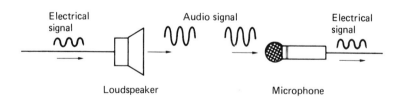

Fig. 5.1 Loudspeaker and microphone

Both of these devices convert one form of energy into another and this is all a transducer is.

34

Commonly used sensors/transducers

(a) Thermistor

Temperature measuring or indicating devices often incorporate thermistors (Fig. 5.2). These devices change in electrical resistance as the temperature changes. The temperature range, however, depends on the type of thermistor and varies a great deal from one to another. These devices are discussed in more detail later.

Fig. 5.2 General purpose thermistor

(b) Ultrasonic transducers

Miniature ultrasonic transmitters and receivers (Fig. 5.3) are useful for remote switching applications. A typical use is the remote control of TV receivers. Operation is generally at about 40 kHz which is well above the highest frequency that the human ear can detect of 20 kHz.

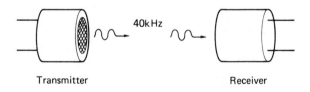

Transmitter   Receiver

Fig. 5.3 Ultrasonic transmitter and receiver

(c) Gas sensors

These are designed to detect flammable gases such as propane, butane, natural gas, etc.

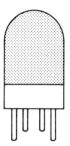

Fig. 5.4 Gas sensor

(d) Strain gauges

Strain is an engineering term used to describe a proportional change in dimensions. When a material (see Fig. 5.5) is loaded its dimensions change and strain is calculated by the formula:

$$\text{Strain} = \frac{\text{change in dimension } x}{\text{original dimension } l}$$

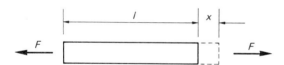

Fig. 5.5 Material under load

The engineer must take this very important factor into consideration when designing structures, frameworks, bridges, aircraft wings, etc. To facilitate convenient detection of strain, tiny devices known as strain gauges were developed. Strain gauges (Fig. 5.6) are taped onto the structure and connected to an electrical circuit known as a 'Wheatstone bridge'. As the structure deforms, the strain gauge's dimensions also change and this produces an increase (or decrease) in its electrical resistance. This alters the conditions of the 'Wheatstone bridge' circuit and provides the engineer with an indication of strain.

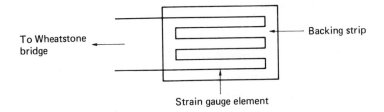

To Wheatstone bridge

Backing strip

Strain gauge element

Fig. 5.6 Typical strain gauge

(e) Light Dependent Resistors (LDRs)

The resistance of these components changes with light intensity. They have a clear window at the top so that light can hit the sensitive element (Fig. 5.7). The resistance of LDRs is high under dark conditions and falls with increase in light intensity. This component is also discussed in more detail later.

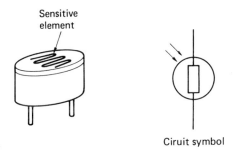

Sensitive element

Ciruit symbol

Fig. 5.7 Light dependent resistor

Connecting sensors to the microcomputer

Each of these sensors and many more, can be linked to the microcomputer in some way so that particular quantities can be monitored. Unfortunately these sensors are analogue and the signals from them would be meaningless to a digital device like the micro. To obtain a meaningful and useable signal for the micro the output from the sensors is fed into an **analogue to digital converter** and the digital output from this converter is fed to the micro user port. A block diagram of this system is given in Fig. 5.8.

Sensor → ADC → Micro

Fig. 5.8 Analogue to digital converter

Analogue and digital signals are discussed in detail later.

Examining the contents of memory

Incoming signals from sensors (via an ADC) are received at the user port and they set bits in the data register to indicate that an input is received. In monitoring applications it is necessary to examine the DR to observe (and perhaps respond to) incoming signals. It is necessary therefore to 'read' the contents of this register to see if a bit has been set. Before going on to this, let us see how we can read or examine the contents of the computer's memory.

Examining the memory is a simple process; understanding what we find is not so easy. Many machines use a PEEK instruction although the BBC uses a ? instruction. We can 'PEEK' into memory as follows:

PROG. 5.1 Examination of the contents of memory location 800

```
       BBC              COMMODORE 64                PET

10 PRINT ?800   10 PRINT PEEK(800)   10 PRINT PEEK(800)

20 END          20 END               20 END
```

In actual fact no line numbers are required and we could obtain the same function by:

PRINT PEEK (800) |RETURN| for the PET and Commodore 64

BBC equivalent is:

PRINT?800

The result displayed is a value between 0 and 255. This is because all the data the machine deals with are in bytes, i.e. binary numbers eight bits long, and so only numbers between 0 and 255 decimal can be handled. Try looking into other locations; the following program allows the operator to input the desired location number.

PROG. 5.2

```
BBC                COMMODORE 64              PET

10 INPUT A         10 INPUT A          10 INPUT A

20 PRINT ?A        20 PRINT PEEK(A)    20 PRINT PEEK(A)

30 GOTO 10         30 GOTO 10          30 GOTO 10
```

'A' being the memory location to be examined or read.

For users of the PET Commodore 64 there is one location on each which comes in very useful when using assembly language for control. These locations are 151(PET) and 197(Commodore 64). The location stores the key depressed value. Run the following program and press any key and observe the display change as each is pressed.

PROG. 5.3

```
        PET                          COMMODORE 64
10  PRINT PEEK(151)         10  PRINT PEEK(197)
20  GOTO 10                 20  GOTO 10
```

Data Direction Register (DDR)

As previously stated the bits in the DDR determine input or output conditions. All 1s mean all lines are outputs, and all 0s mean all lines are inputs, and it is possible to examine its state in just the same way as we did for the other memory locations. The instruction is:

```
     BBC          COMMODORE 64              PET
PRINT?65122    PRINT PEEK(56579)   PRINT PEEK(59459)
```

Program 5.4 sets all the lines at the user port as inputs and then prints the value in the DDR to prove they are all 0s (inputs).

PROG. 5.4 Setting and inspecting the DDR

```
      BBC              COMMODORE 64             PET
10 ?65122=0         10 POKE 56579,0     10 POKE 59459,0
20 PRINT ?65122     20 PRINT PEEK(56579) 20 PRINT PEEK(59459)
30 END              30 END               30 END
```

Data register (input/output register)

Examination of the data direction register during control sequences is not necessary, fortunately, because once the control lines are set they remain set throughout.

Incoming signals set bits in the data register (DR) and it is this register, therefore, which must be periodically or systematically inspected. One problem arises when using the DR and this is that a '1' is set in this register if *either* an input is being received *or* the micro is emitting a signal. To illustrate this more clearly, imagine that four lines are set as inputs and four as outputs, and take the extreme case that all four input lines are actually all receiving input signals simultaneously, and four outputs are being sent at the same time. If we now PEEKed into the DR we would obtain a value of 255 (11111111 binary) and from this it is impossible to tell which are outputs and which are inputs. Fortunately this problem is not great and with a little experience it can be overcome.

Practical situations involving input signals are given in several chapters, but a simple circuit is given here which can be used to simulate inputs.

Input simulator

The output circuit board described in Chapter 4 can be modified to provide an integrated input/output board. A simple resistor and switch arrangement can be connected between the base of each transistor and ground (0V) as indicated in Fig. 5.9. When the switch is open P0 is at logic 1 and the LED is illuminated — this simulates an input. When the switch is closed P0 is grounded, the transistor is switched off and the LED is off indicating no input. A small resistor (1kΩ) can be placed at position × if desired.

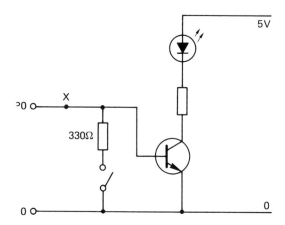

Fig. 5.0 Circuit for input simulator

Because eight switches are required (one for each transistor base) it is suggested that miniature single pole (s.p.) switches are used (see Fig. 5.10).

The circuit does not actually produce an input (simulation of an input only is obtained) but it is extremely useful for practising handling and responding to input signals. The main thing to recognise is the problem that if only one line is being used as an input at the user port and all the others are disconnected they will be at logic 1, thereby indicating some inputs when in actual fact there are

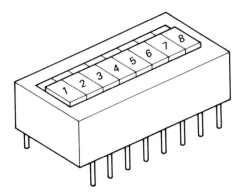

Fig. 5.10 Miniature single pole switches

none. The skilled programmer soon becomes able to ignore the distracting effect of the other lines.

Program 5.5 sets all the input/output lines as INPUTS and then reads (and prints on screen) continuously the value found in the DR. Switches can then be closed to simulate no input or opened to simulate an input.

PROG. 5.5 Monitoring input signals

BBC	PET	COMMODORE64
10 ?65122=0	10 POKE 59459,0	10 POKE 56579,0
20 PRINT ?65120	20 PRINT PEEK (59471)	20 PRINT PEEK (56577)
30 GOTO 20	30 GOTO 20	30 GOTO 20

Initially, if all switches are closed, the reading will be 0 and as each switch is opened the reading is the decimal value of the particular bit weight. That is 1, 2, 4, 8, 16, 32, 64, 128 or combinations of these.

Responding to inputs

Control software often involves producing some response to input signals and it is possible to write simple programs for the input simulator to provide an introduction to this vital area.

PROG. 5.6 Responding to an input

BBC	COMMODORE 64	PET
10 ?65122=127	10 POKE 56579,127	10 POKE 59459,127
20 ?65120=0	20 POKE 56577,0	20 POKE 59471,0
30 IF ?65120=128 THEN 50	30 IF PEEK(56577)=128 THEN 50	30 IF PEEK(59471)=128 THEN 50
40 GOTO 20	40 GOTO 30	40 GOTO 30
50 ?65120=127	50 POKE 56577,127	50 POKE 59471,127
60 END	60 END	60 END

Line 10 sets 7 lines output and 1 input (P7).
Line 20 switches off all LEDs.
Line 30 tests for an input on P7 (128).
Line 40 provides continuous testing loop.

Line 50 switches on seven LEDs on receiving an input on P7.
Line 60 ends program.

Line 30 in program 5.6 read—IF PEEK (59471) = 128 THEN 50. This was used to test for an incoming signal. But suppose in a real situation several devices were producing inputs. Then at times two or even several inputs could be received simultaneously. The value in the DR may then never equal 128 and so a real problem arises. What is required is an instruction which allows inspection of one line only, i.e. the testing of one particular bit. This can be done using LOGICAL instructions and because these are extremely important in control applications a complete chapter has been devoted to them. However, if you are itching to solve the problem now, then line 30 should read:
IF PEEK (56577) AND 128 THEN 50 (Commodore 64)
BBCB and PET programs can be modified similarly. With this instruction no matter how many inputs there are only P7 being inspected for an input.

6

Stepper motors

Stepper motors, unlike conventional d.c. motors, do not provide a smooth, continuous motion. The output shaft of a stepper motor rotates by incrementing in discrete steps—rather like the second hand on a clock. The speed of the second hand on a clock is fixed of course but the stepper motor speed can be varied quite easily, as will be seen later. A typical stepper motor is shown in Fig. 6.1.

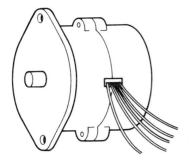

Fig. 6.1 A typical stepper motor

The second hand of a clock completes one revolution by stepping sixty times; each step is therefore 6°. Typical inexpensive stepper motors have a resolution of 7.5°, that is forty-eight steps/revolution. More accurate and expensive types have a resolution of 1.8°, (200 steps/revolution). However, it is possible with these to achieve 3.75° and 0.9° resolution respectively by using a technique known as 'single-dual phase excitation'—not discussed in this text.

Stepper motor vs d.c. motor

When precise positional control (for example with robot arms and CNC lathes) is required with d.c. motors, some form of feedback mechanism must be employed so that the position is known. A feedback signal is sent to the controller (a microprocessor perhaps) and so adjustments can be made until the required position is achieved. D.c. servo systems are often employed in such positional control systems. However such systems, with this fairly complex electronic circuitry and relatively high cost, are not that relevant to the beginner and so are not described in this introductory text.

Stepper motors have the distinct advantage that a continuous feedback is not essential. Positional control is achieved simply by stepping the motor through the required angle. Naturally the more steps/revolution offered by the motor the greater is the precision of positional control.

Principle of the stepper motor

The motor shaft is made to rotate by the attraction of unlike magnetic poles. An illustration of how rotation can be produced by such attraction is given in Fig. 6.2. A permanent magnet is brought near to a compass needle. If the magnet is now rotated about the axis of the compass the needle will follow and try to align itself with the magnet.

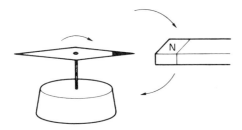

Fig. 6.2 Producing rotation using magnetic fields

In actual fact a stepper motor spindle (the rotor arrangement) houses numerous magnetic poles and instead of a rotating permanent magnet several electromagnets are located around the motor body (the stator). Figure 6.3 shows a section of a motor and just coil 1 is wired for clarity, but as can be seen every fourth coil is connected together. The number of coils on the stator is four times the number of magnetic poles on the rotor and this number determines the step angle. Coils 1, 2, 3 and 4 are all connected together at one end and so there are five wires out of the motor. Current must be fed to these coils in a set sequence to produce a continual rotation.

A hardware system or a software system or a combination of the two can be used to produce the required coil energising sequence. Here a software system only is described.

The magnetic poles 'P' in Fig. 6.3 are at present between coils 1 and 2 and as they are both energised and attracting the poles the rotor will be held at that position. If current to coil 1 is now switched off and coil 3 is energised the rotor will move clockwise until it is between coils 2 and 3, as shown in Fig. 6.4, and it will now be held in this position. A single step has taken place, and to step the motor again the current to coil 2 is removed and coil 4 energised.

The energising sequence is therefore, coils 1 and 2, then 2 and 3, then 3 and 4 and back to coil 1 with 4 and 1. Naturally to keep the motor running the

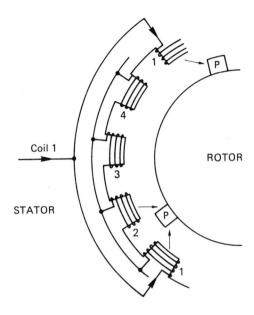

Fig. 6.3 A section of a stepper motor

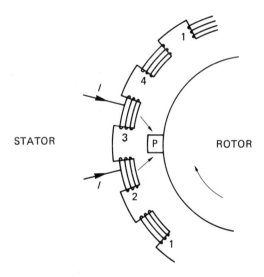

Fig. 6.4

sequence is repeated. To run the motor in the opposite direction the sequence is simply reversed.

Table 6.1 shows the sequence that must be generated at the user port, assuming that coils 1, 2, 3 and 4 are connected to lines P0, P1, P2 and P3 respectively (via interface circuitry of course).

Table 6.1 Coil energising sequence

Step number	Coil 4 P3(8)	Coil 3 P2(4)	Coil 2 P1(2)	Coil 1 P0(1)	Value 'poked' to user port
1	0	0	1	1	3
2	0	1	1	0	6
3	1	1	0	0	12
4	1	0	0	1	9

Stepper motor interface

Ready-to-use TTL-interfaced stepper motors can be purchased from many micro-electronic equipment manufacturers, and although much more expensive than home built models, they are ideal for the beginner.

The motors themselves can be obtained separately in a wide range of sizes and stepping angles but then the problem of interfacing arises. Figure 6.5 is a general interface circuit but the actual components, in terms of size and value, depend on the size of the stepper motor to be driven. The operation of the circuit is that of transistor switching. Resistors R1/R4 are chosen to provide suitable transistor switching conditions and the four diodes D1/D4 provide protection against back e.m.f.s.

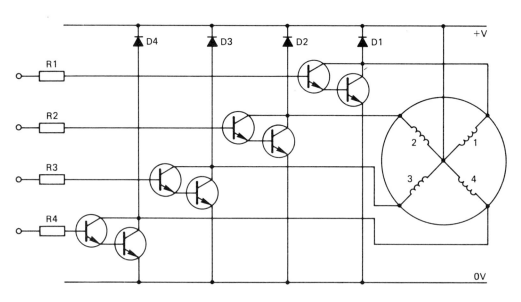

Fig. 6.5 General stepper motor interface circuit

Alternatively a darlington driver chip (and accompanying relays if necessary) can be utilised. Unfortunately only two stepper motors can be controlled from the user port. However on the market now are 'stepper motor interface/driver' chips and with these one simply controls the direction of rotation and the number of steps. The 'chip' sorts out the sequencing problems.

Program 6.1 runs the stepper motor continuously in one direction. It is rather unsatisfactory, although it works after a fashion, in that the micro computer will generate pulses at a rate such that the motor may be unable to respond due to the inertia of its own rotor.

PROG. 6.1 Basic stepper motor control

BBC	COMMODORE 64	PET
10 ?65122= 255	10 POKE56579,255	10 POKE59459,255
20 ?65120=3	20 POKE56577,3	20 POKE59471,3
30 ?65120=6	30 POKE56577,6	30 POKE59471,6
40 ?65120=12	40 POKE56577,12	40 POKE59471,12
50 ?65120=9	50 POKE56577,9	50 POKE59471,9
60 GOTO 20	60 GOTO 20	60 GOTO 20

Line 10 sets all the control lines as outputs, although only four are required.
Lines 20, 30, 40 and 50 produce the desired sequence of pulses.
Line 60 provides continuous running of the motor.

Program 6.2 is more satisfactory. It allows the operator to input a desired value which will be used in a delay subroutine to control the speed of the motor.

PROG. 6.2 Motor speed control

```
                    BBC
10 INPUT"DELAY VALUE",D
20 ?65122=255
30 ?65120= 3
40 GOSUB 500
50 ?65120= 6
60 GOSUB 500
70 ?65120= 12
80 GOSUB 500
90 ?65120= 9
100 GOSUB 500
110 GOTO 30
500 REM DELAY
510 FOR I=0 TO D
520 NEXT I
530 RETURN
```

```
        COMMODORE 64                    PET
10 INPUT"DELAY VALUE",D     10 INPUT"DELAY VALUE",D
20 POKE56579,255            20 POKE59459,255
30 POKE56577,3              30 POKE59471,3
40 GOSUB 500                40 GOSUB 500
50 POKE56577,6              50 POKE59471,6
60 GOSUB 500                60 GOSUB 500
70 POKE56577,12             70 POKE59471,12
80 GOSUB 500                80 GOSUB 500
90 POKE56577,9              90 POKE59471,9
100 GOSUB 500               100 GOSUB 500
110 GOTO 30                 110 GOTO 30
500 REM DELAY               500 REM DELAY.
510 FOR I=0 TO D            510 FOR I=0 TO D
520 NEXT I                  520 NEXT I
530 RETURN                  530 RETURN
```

Line 10 allows the operator to control the speed of rotation by the length of the
    delay 'D'.
Line 20 sets all the control lines as outputs.
Lines 30, 50, 70 and 90 are the required pulse sequence.
Lines 40, 60, 80 and 100 bring in the delay routine housed in lines 510 and 520.

It is a fairly simple task to govern the number of complete (or parts of)
revolutions made by the motor because each step angle is known. Program 6.3
assumes a stepping angle of 7.5° and so the pulse sequence (3, 6, 12, 9) must be
repeated 12 times to produce each revolution.

PROG. 6.3 Positional control of a stepper motor

BBC

```
 10 INPUT"NUMBER OF REVS",
 20 INPUT"DELAY VALUE",D
 30 N=0
 40 ?65122=255
 50 ?65120= 3
 60 GOSUB 500
 70 ?65120= 6
 80 GOSUB 500
 90 ?65120= 12
100 GOSUB 500
110 ?65120= 9
120 GOSUB 500
130 N=N+1
140 S=(N/12)
150 IF S=R THEN 170
160 GOTO 50
170 END
500 REM DELAY
510 FOR I=0 TO D
520 NEXT I
530 RETURN
```

PET

```
 10 INPUT"NUMBER OF REVS",R
 20 INPUT"DELAY VALUE",D
 30 N=0
 40 POKE59459,255
 50 POKE59471,3
 60 GOSUB 500
 70 POKE59471,6
 80 GOSUB 500
 90 POKE59471,12
100 GOSUB 500
110 POKE59471,9
120 GOSUB 500
130 N=N+1
140 S=(N/12)
150 IF S=R THEN 170
160 GOTO 50
170 END
500 REM DELAY
510 FOR I=0 TO D
520 NEXT I
530 RETURN
```

COMMODORE 64

```
 10 INPUT"NUMBER OF REVS",R
 20 INPUT"DELAY VALUE",D
 30 N=0
 40 POKE56579,255
 50 POKE56577,3
 60 GOSUB 500
 70 POKE56577,6
 80 GOSUB 500
 90 POKE56577,12
100 GOSUB 500
110 POKE56577,9
120 GOSUB 500
130 N=N+1
140 S=(N/12)
150 IF S=R THEN 170
160 GOTO 50
170 END
500 REM DELAY
510 FOR I=0 TO D
520 NEXT I
530 RETURN
```

The only new lines in this program are:
Line 10 asks for the number of revolutions to be completed.
Line 30 sets a counter 'N' to zero.
Line 130 increments the counter by one each time the energising sequence is
           completed (i.e. angle of 30°).
Line 140 calculates how many revolutions have been completed.
Line 150 tests if the number of revolutions made is equal to the number required
           and if it is then line 170 stops the rotation.

Exercise programs (solutions on p. 163)

Write the following programs for the stepper motor. The solutions assume a 7.5°
angle of step.
6.4 Control the number of revolutions, the speed and the direction of the motor.
6.5 Control the motor from lines P4-P7.

Exercise program (without solution)

Write a program to determine the speed of rotation by using the microcomputers own timing mechanism.

7

D.C. motor with revolution counter

This is one of the simplest projects discussed in this text and in fact leads to some of the most interesting and enlightening control and monitoring exercises for the beginner. It involves the control of a small d.c. motor whilst simultaneously counting the number of revolutions made. The programmer can also make effective use of the microcomputer's internal timing mechanism for motor speed determination. Figure 7.1 shows the block diagram of the system.

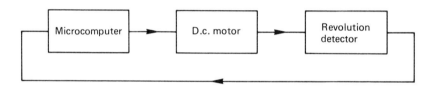

Fig. 7.1 Block diagram of the d.c. motor revolution counting system

Simple d.c. motor

The arrangement of the components making up a simple d.c. motor is given in Figure 7.2 and the principle is as follows:

current from the battery flows around the circuit in the direction of the arrows. As this current flows through the copper conductor (A to B) it sets up a magnetic field around it. This field in turn interacts with the field of the permanent magnets. Interaction of these fields produces a force (hence motion) on the conductor. Direction of motion is anticlockwise in this instance, but this can be reversed by simply reversing the battery connections. The actual direction of rotation can be found using Fleming's Left Hand Rule; however, this is beyond the scope of this text. Motor speed is governed by the current and hence the battery voltage, and, because voltage can generally easily be varied, the speed of the motor can be easily controlled.

Small d.c. motors, like the one illustrated in Figure 7.3, are relatively cheap and suitable for this project.

As described in Chapter 3 switching motors on and off produces induced (or back) e.m.f.s and so to protect circuitry a diode (see Fig 7.4) should be

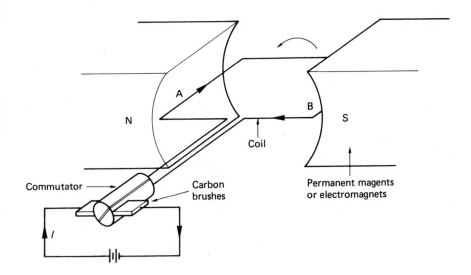

Fig. 7.2 A simple d.c. motor

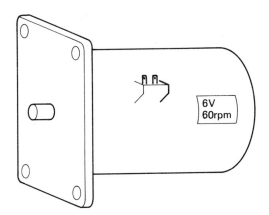

Fig. 7.3 A small, inexpensive motor

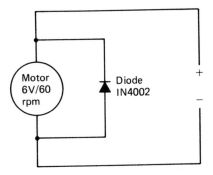

Fig. 7.4 Diode protection

connected across the motor. This means that the direction of motor rotation cannot be reversed unless the diode is also reversed.

Revolution counting

There are several different components which could be employed to provide 'revolution detection'. The one described here, shown in Fig 7.5(i), uses an opto-switching mechanism.

A slotted disc, Fig 7.5(ii) which can be fixed to the motor spindle is required. Aluminium is the ideal material although not essential.

Figure 7.5(iii) is a closer look at the opto-switching device, and shows that it consists of an infra-red emitting LED and at the opposite side is a phototransistor. Light from the LED switches the phototransistor on, and so as the slotted disc moves between the two components the phototransistor will be 'off' except when the slot is encountered. At this instant a signal will be produced and this is received by the microcomputer.

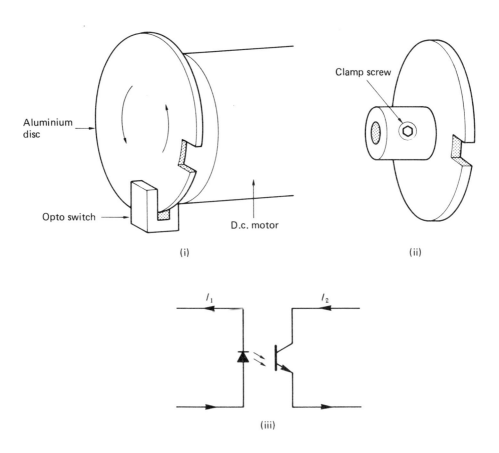

Fig. 7.5 Opto-switching

Circuit diagrams and assemblies

Opto-switch circuit

Figure 7.6 is the complete circuit diagram for the opto-switching arrangement. A 5V supply provides a continuous current to the opto's LED via the 150Ω resistance. When the rotating disc is blocking the light, and the phototransistor is not conducting electricity, current from the supply flows through the 22 kΩ resistance to switch on the BC109 transistor. This directs current, flowing through the 510Ω resistance to ground (0V) and so the potential at 'A' is low and no input to the user port is obtained.

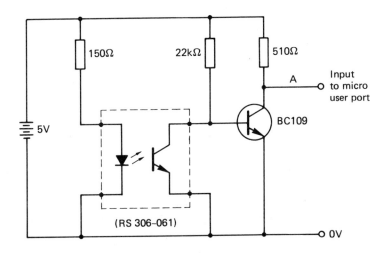

Fig. 7.6 Opto-switch circuit

At the instant the slot is encountered, and light hits the phototransistor, the reverse of the previous conditions occur. That is, the BC109 is switched off and so the potential at 'A' is high (5V) and an input signal to the user port is produced.

Motor/opto-switch assembly

The finished arrangement could look similar to that shown in Fig. 7.7. Careful mounting and rigidity is required for the motor and accurate machining and location of the slotted disc is necessary. However, in one aspect of the disc design accuracy is not essential and it may be surprising to find that this is in fact the width of the slot itself. The reason for this seeming anomaly is simply that the software will, or should, be such as to cater for any width of slot.

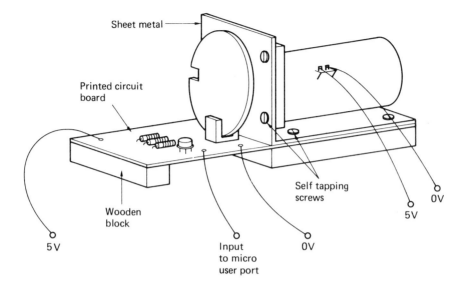

Fig. 7.7 Completed assembly

General wiring diagram

If you have decided to produce the 'output circuit' described in Chapter 3 then the control of the motor can be directly through this. Alternatively the motor could be driven through a pair of transistors as was also previously described. If the latter is adopted, careful selection of transistors to suit the motor must be made. Assuming the first option has been chosen then the wiring diagram is as shown in Fig. 7.8. Note again that all the 0V lines are connected together.

PROG. 7.1 Control of d.c. motor

BBC	COMMODORE 64	PET
10 ?65122=127	10 POKE 56579,127	10 POKE 59459,127
20 ?65120=1	20 POKE 56577,1	20 POKE 59471,1
30 FOR I=0 TO 8000	30 FOR I=0 TO 8000	30 FOR I=0 TO 8000
40 NEXT I	40 NEXT I	40 NEXT I
50 ?65120=0	50 POKE 56577,0	50 POKE 59471,0
60 END	60 END	60 END

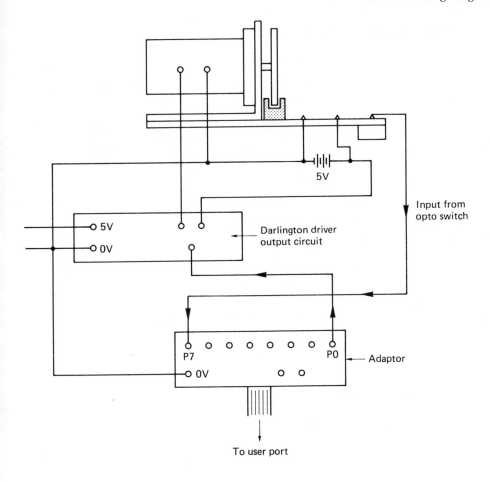

Fig. 7.8 Motor/opto-switch wiring diagram — using the 'output circuit'

Line 10 sets seven control lines as output and one as an input line. The input line P7 (128) will be used later to receive signals from the opto-switch.
Line 20 switches the motor (connected to P0) on.
Lines 30 and 40 provide a short time delay.
Line 50 switches the motor off.

All that is achieved here is simply switching the motor on for a short period of time. The following program switches the motor on for a reasonable length of time and simultaneously displays on the screen the value found in the data register. Any input signals will therefore appear as numbers between 0 and 255 depending on which line they appear. We are only expecting an input on line P7; therefore a value of 128 is the expected value to be displayed on receiving a signal.

PROG. 7.2 Monitoring the opto-switch

BBC	COMMODORE 64	PET
10 ?65122=127	10 POKE56579,127	10 POKE59459,127
20 ?65120=1	20 POKE56577,1	20 POKE59471,1
30 FOR I=O TO 30000	30 FOR I=O TO 30000	30 FOR I=O TO 30000
40 PRINT?65120	40 PRINTPEEK(56577)	40 PRINTPEEK(59471)
50 NEXT I	50 NEXT I	50 NEXT I
60 ?65120=0	60 POKE56577,0	60 POKE59471,0
70 END	70 END	70 END

Line 10 sets seven output lines and one input line.
Line 20 switches on the motor.
Lines 30/50 are time delay.
Line 40 prints, on the screen, the value found in the data register and so indicates
    input signals.
Line 60 switches the motor off.
On running the program the screen displays something like this:

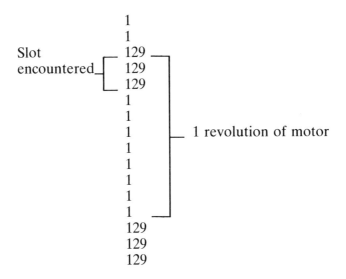

Let us analyse the displayed values.
    As the disc is rotating with the motor spindle a value of '1' is displayed
until the slot is encountered by the opto-switch and then a '129' is displayed.
The '1' is the bit set in the data register for switching the motor on and so
when the slot is encountered a signal on P7 (weighted as 128) is also received.
    From the results displayed it is obvious that as the slot passes the
opto-switch the microcomputer has time to read the data register three times,
in this case, as can be seen by the succession of 129s. The next problem
therefore is to design software so that the displayed information allows the
number of revolutions to be recorded.

BBC

```
 10 INPUT"REVS";R
 20 ?65122=127
 30 ?65120=1
 40 N=0
 50 IF ?65120=1 THEN 50
 60 IF ?65120=129 THEN 60
 70 N=N+1
 80 PRINT N;"REVS"
 90 IF N=R THEN 110
100 GOTO 50
110 ?65120=0
120 END
```

COMMODORE 64

```
 10 INPUT"REVS";R
 20 POKE56579,127
 30 POKE56577,1
 40 N=0
 50 IF PEEK(56577)=1 THEN 50
 60 IF PEEK(56577)=129 THEN 60
 70 N=N+1
 80 PRINT N;"REVS"
 90 IF N=R THEN 110
100 GOTO 50
110 POKE56577,0
120 END
```

PET

```
 10 INPUT"REVS";R
 20 POKE59459,127
 30 POKE59471,1
 40 N=0
 50 IF PEEK(59471)=1 THEN 50
 60 IF PEEK(59471)=129 THEN 60
 70 N=N+1
 80 PRINT N;"REVS"
 90 IF N=R THEN 110
100 GOTO 50
110 POKE59471,0
120 END
```

Line 10 allows the operator to input the number of revolutions he or she wishes the motor to complete.

Line 20 sets 7 output lines and 1 input line.

Line 30 starts the motor.

Line 40 zeros the revolution counter.

Line 50 says 'if there is no input from the opto-switch wait at line 50'.

Line 60 says 'if there is an input from the opto-switch wait at line 60'.

Line 70 adds '1' to the revolution count.

Line 80 displays the number of revolutions completed.

Line 90 tests to see if revolution count equals input value R.

Line 100 returns to line 50 if N does not equal R.

Line 110 stops the motor if N equals R.

Lines 50 and 60 use an equal sign ($=1$; $=129$) but a more satisfactory way is to use logical instructions. These, however, are discussed later.

The VDU displays:

```
REVS? 8
1 REVS
2 REVS
3 REVS
  .
  .
  .
. ETC
```

Lines 50 and 60 are the critical lines and both these hold the program sequence until a change in the data register is observed. This means that no matter how large the slot in the disc is made the result is the same. Try altering the speed (by voltage adjustment) of the motor and run the program again.

Exercise program (solution on p. 164)

We are now in a position to make full use of the microcomputer's internal timing mechanism for determination of the motor's speed. This is calculated mathematically by:

$$speed = \frac{number\ of\ revolutions}{time\ in\ minutes}$$

This formula must obviously be incorporated in your program and because our previous program records the number of revolutions made then the time factor is the only extra. In the solution I have called the start time 'A', the finish time 'B' and the difference (or time taken) 'C'.

Such revolution counting or speed determination arrangements have many real applications, whether they are fitted to motor output spindles or any other rotating shaft, such as gearbox shafts, and monitoring such devices using a microprocessor is becoming more and more widespread.

The main thing to realise about controlling the d.c. motor is that precise angular control is impossible with the system above and a much more elaborate and expensive system would be required. This is why stepper motors are becoming so popular these days. I hope this project, with all its limitations, has given some useful programing exercises as well as an insight into motor control and revolution counting.

8

Digital to analogue converters

Digital signals and devices

Computers understand and operate with digital signals only, as do calculators. Such signals have only two states or levels; either on or off or a voltage high or a voltage low. A light switch is another common digital device and the two states it can be in, 'on' or 'off', can be represented by a '1' or '0' respectively. Figure 8.1 illustrates a typical digital signal.

Fig. 8.1 A typical digital signal

Analogue signals and devices

In the real world many devices are non-digital — or analogue as they are generally known. An analogue device is one in which a continuously changing input produces a continuously changing output.

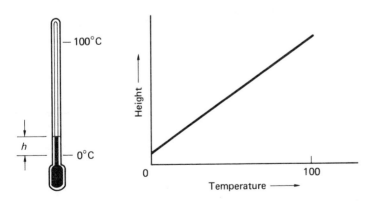

Fig. 8.2 An analogue device

59

A typical analogue device is a thermometer. The height of the mercury (output) is proportional to the ambient temperature (input). Figure 8.2 shows a graph of 'height of mercury' plotted against temperature. The graph is a continuous line thus indicating an analogue device, and from this graph it is apparent that there are an infinite number of possible input values (even in the range 0–100°C) and an infinite number of possible output values—this is a 'test' for an analogue device.

Digital to Analogue Converter — DAC

Often the output signal from a digital system, such as a microcomputer, has to be converted into analogue form. The output may be required to drive say, a graph plotter or control the speed of a d.c. motor. The multiple bit digital output is converted into an analogue voltage by a DAC and this voltage is proportional to the binary value of the digital signal fed into the DAC.

To understand this more clearly let us examine the input and corresponding output from the DAC in Fig. 8.3. It is an eight bit device, that is, there are eight lines into it (P0-P7) from the microcomputer. The signals from the micro computer can vary between:

$$00000000 \quad \text{binary} \quad (0 \text{ denary}) \quad \text{to}$$
$$11111111 \quad \text{binary} \quad (255 \text{ denary})$$

If the analogue output from the DAC can vary between say 0 and 2.55 volt, then hopefully it is not too difficult to comprehend that an input to the DAC of 0 denary means the output is 0 volt, and likewise an input of 255 denary produces a voltage of 2.55 volt. The relationship can be found as follows:

$$\frac{2.55 \text{ volt}}{255 \text{ denary}} \quad = \quad 0.01 \text{ volt/denary unit}$$

So if the output from the microcomputer is 10000000 binary (128 denary) a voltage of 1.28 volt is generated by the DAC.

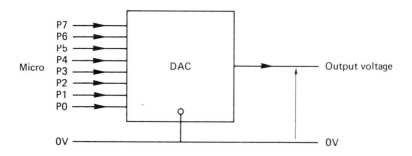

Fig. 8.3 A typical DAC

The accuracy of an 8 bit DAC is not great; obviously a 12 bit device offers much greater accuracy, but an 8 bit is quite adequate for general use as will be seen later.

Principle of operation

Figure 8.4 shows a theoretical 3 bit DAC. The incoming signals close electronic switches and these connect voltage levels to the operational amplifier which sums them and outputs the analogue voltage.

A truth table for the 3 bit DAC is given below:

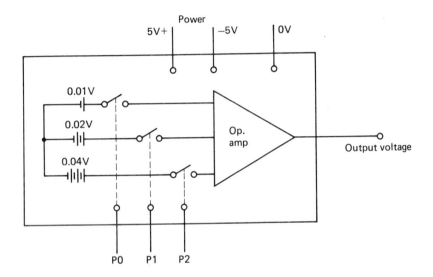

Fig. 8.4 Principle of a DAC

Table 8.1

Input			Output
P_2	P_1	P_0	Volts
0	0	0	0
0	0	1	0.01
0	1	0	0.02
0	1	1	0.03
1	0	0	0.04
1	0	1	0.05
1	1	0	0.06
1	1	1	0.07

With our simple 3 bit arrangement it is possible to produce 8 different outputs and with an 8 bit DAC 256 voltage levels (including 0V) are possible.

There are many converters on the market at a wide range of prices and many offer negative as well as positive voltage outputs. However, generally the enthusiast will not require this added sophistication.

One can buy a combined DAC and ADC (analogue to digital converter) on one board from certain manufacturers which is a great advantage and when using such devices all that is necessary is for the operator to select DAC or ADC by connecting a 5V here or a 0V there. It is also possible to purchase these converters in kit form, or obtain circuit diagrams from magazines, but unless you are a keen electronics enthusiast I would advise obtaining a 'working model' by paying a little more.

DAC drivers

Although the DAC board allows the control of voltage over a specified range by computer programming, the actual current that it can supply is very small and so limits its practical use. High power transistor driver circuits can be obtained or made which can be added on to the DAC, so that much greater power is available for the control application.

However, for all the following programs in this chapter no driver is required. The only additional equipment required is a simple d.c. voltmeter, capable of indicating the maximum DAC output voltage. But, if you have access to an oscilloscope or even better an X, Y plotter more interesting observations can be made.

Wiring diagram

Figure 8.5 shows how a typical DAC would be wired ready for use. Note that a negative voltage from a power source is required.

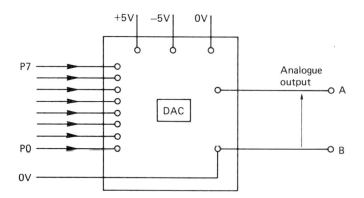

Fig. 8.5 A typical wiring diagram for a DAC

Programming the microcomputer to control d.c. voltage via a DAC is quite a rewarding experience even if nothing is actually being driven. In order to observe the output voltage changes a d.c. voltmeter must be connected between

A and B in Fig. 8.5. Alternatively an oscilloscope or an X, Y graph plotter can be connected between A and B (Fig. 8.6).

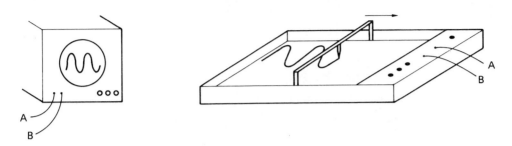

Fig. 8.6 Oscilloscope and X, Y plotter in use with DAC

The first program in this section allows the direct monitoring of the output voltage. In this case the use of the voltmeter is preferred.

PROG. 8.1 Voltage control

PET

```
10 POKE59459,255
20 INPUT"INPUT VALUE";V
30 POKE59471,V
40 GOTO 20
```

COMMODORE 64

```
10 POKE56579,255
20 INPUT"INPUT VALUE";V
30 POKE56577,V
40 GOTO 20
```

BBC

```
10 ?65122=255
20 INPUT"INPUT VALUE",V
30 ?65120=V
40 GOTO 20
```

Line 10 sets all control lines as outputs.
Line 20 allows the operator to chose a value to be output.
Line 30 produces the output signal which is then converted by the DAC.
Line 40 starts sequence again.

With this simple program it is possible to input numbers between 0 and 255 and observe the corresponding voltmeter readings.

PROG. 8.2 Generating a sawtooth waveform

PET

```
10 REM SAWTOOTH WAVE
20 POKE59459,255
30 FOR X=0 TO 255
40 POKE59471,X
50 NEXT X
60 GOTO 30
```

COMMODORE 64

```
10 REM SAWTOOTH WAVE
20 POKE56579,255
30 FOR X=0 TO 255
40 POKE56577,X
50 NEXT X
60 GOTO 30
```

BBC

```
10 REM SAWTOOTH WAVE
20 ?65122=255
30 FOR X=0 TO 255
40 ?65120=X
50 NEXT X
60 GOTO 30
```

The sawtooth is produced by lines 30, 40 and 50 by simply incrementing the number that is output from the user port and then returning to start the sequence again when 255 is reached.

If an oscilloscope is used to observe the sawtooth it will require a slowtime base setting thus illustrating how slow BASIC really is. If an X, Y plotter is used the program speed should be satisfactory. If not a delay can be introduced into the program.

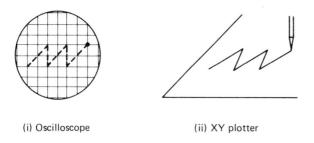

(i) Oscilloscope   (ii) XY plotter

Fig. 8.7 Display of the sawtooth waveform

Microcomputers are ideally suited, if this is not an understatement, to manipulating mathematical equations and functions and so using a DAC one can show these functions on an X, Y plotter. For anyone involved in teaching mathematics or science this facility is invaluable. All that is necessary is for the function to be introduced into the software in such a way that the calculated function does not exceed 255, by introducing a scale factor.

Figure 8.8 shows an exponential function produced on an X, Y plotter.

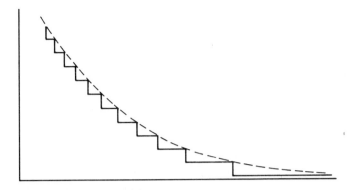

Fig. 8.8 A generated exponential function

Exercise progams (solutions on p. 165)

Write the following programs for use with a DAC.
8.3 Generate a square wave.
8.4 Generate a sinewave.

9

Logical operators

Logical AND, OR and NOT instructions are essential in many control situations. Often particular control lines have to be set at logic 1 or logic 0 to switch on or switch off particular devices without affecting any of the other input/output lines. This requires the setting or clearing of the corresponding bits in the input/output register (DR), and logical instructions prove ideal for this purpose. These instructions also allow us to examine the state of particular bits in the DR and this is necessary in many control applications if an input signal on a particular line is to trigger off some control sequence.

To understand these instructions fully we must examine logic gates. Logic gates are electronic circuits performing logical actions. Although such gates are not directly relevant they do provide a simple analysis of AND and ORs.

Logic gates

AND gate

A simple circuit which illustrates the operation of an AND gate is shown in Fig. 9.1. The lamp 'S' lights only when switches **A AND B** are closed. A list of conditions of the circuit are shown alongside Fig. 9.1.

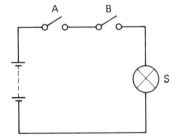

Switches		Lamp
A	B	S
Open	Open	Off
Closed	Open	Off
Open	Closed	Off
Closed	Closed	On

Fig. 9.1 AND circuit

Let us now class a switch open as a '0' and a switch closed as a '1'. Similarly a lamp off as '0' and lamp on as '1'.

66

A new table called a truth table can be drawn up (Table 9.1). From the table it can be seen that the result of ANDING two numbers is '1' only when the numbers are both 1's.

Table 9.1 Truth table

A	B	S
0	0	0
1	0	0
0	1	0
1	1	1

Exercise 9.1 (solutions on p. 166)

AND the following numbers.

   (a) 0101
       1111

   (b) 11100011
       01000010

   (c) 01110011
       00010000

AND the following binary number with denary 2.

   (d) 01111010

OR gate (inclusive)

A simple circuit which illustrates the operation of an OR gate is shown in Fig. 9.2. The lamp 'S' illuminates when switches **A OR B** are closed, and a truth table for this is given in Table 9.2.

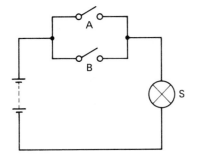

Fig. 9.2 OR circuit

Table 9.2 OR truth table

A	B	S
0	0	0
0	1	1
1	0	1
1	1	1

It can be observed that the result of an OR is '1' when either of the values for A or B are '1'.

Exercise 9.2 (Solutions on p. 166)

    (a)   0101
           1111
           ——

    (b)   11100011
           00010000
           ————

    (c)   01110011
           10010100
           ————

OR the following binary number with denary 4.

    (d)   01111010

NOT gates

Figure 9.3 illustrates the action of a 'NOT gate'. When the switch A is closed, current by-passes the lamp and so it is switched off.

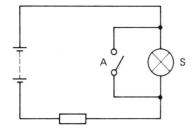

Fig. 9.3 NOT circuit

A truth table (Table 9.3) for such a circuit reveals that the open switch (0) produces an output (1) and vice versa.

Table 9.3 Truth table for a NOT gate

A	S
1	0
0	1

NOT AND (NAND) and NOT OR (NOR) gates

If a NOT gate is added on to the output of either an AND or an OR gate, the final output is simply an inversion of the output from the AND or OR gates. i.e. all the 1s become 0s, and the 0s become 1s.

Logical instructions

Before examining the use of logical instructions for control and monitoring applications, try the following simple exercises.

PROG. 9.1 Use of logical AND (Commodore 64, BBC, PET)

```
10 REM USE OF 'AND'
20 INPUT "SIZE",S
30 INPUT "WEIGHT",W
40 IF S=7 AND W=5 GOTO 60
50 GOTO 20
60 PRINT "CORRECT"
70 END
```

Consider program 9.1. CORRECT is printed if both conditions in line 40 are true. That is, if size is 7 and weight is 5. If either one or both of them is/are false, CORRECT is not printed.

PROG. 9.2 AND/OR instructions (Commodore 64, BBC, PET)

```
10 INPUT B,C
20 PRINT B,C
30 PRINT B AND C
40 PRINT B OR C
50 GOTO 10
```

Program 9.2 allows the operator to input values for B and C which are then ANDed and ORed and the results printed out.

If the values for B and C are both 0s then the result of both logical operations is 0. Likewise an input of 1s for both B and C results in 1s from the logical instructions. If however, the inputs for B and C are 1 and 0 (or 0 and 1) then the result is a 0 for the AND instruction and a 1 for the OR instruction.

Other values could now be used for B and C. Suppose a value for B of 10 and a value for C of 7 are input. The result of the AND would be 2, and of the OR would be 15. To see why this is so the numbers must be converted into binary. The corresponding binary bits are ANDed/ORed in the normal way.

        B   1010   (10 denary)
        C   0111   (7 denary)

        B and C   0010   (2 denary)
        B or C   1111   (15 denary)

The last example illustrates the type of analysis that must be carried out when using logical instructions for control applications.

Testing for a bit set or signal high

To test for a bit high or low, or to set a bit high or low a **mask** is used. The term mask is used to describe the pattern of bits used with the logical instruction to achieve the desired result.

The following instruction can be used to test for a bit set (or signal high) in the data register, so is useful for testing for incoming signals.

                    BBC: Y IF ?65120 AND N THEN X
        COMMODORE 64:   Y IF PEEK(56577)AND N THEN X
                    PET: Y IF PEEK(59471)AND N THEN X

Y is the line number of course, and X is the line number to be executed next if the result is true. N is the pattern of bits (given in denary form) providing this mask.

Example
    The following example illustrates how to test for a signal on P1 (2 denary), using the Commodore 64.

```
10 IF PEEK(56577) AND 2 THEN 200
```

Let us suppose the value in the data register was say 143 denary (10001111 binary), then this is ANDed with 2 below.

data register value	10001111	(143 denary)
mask value	00000010	(2 denary)
result of AND	00000010	(2 denary)

The result of the AND is 2 showing that the signal on line P1 is high and the program sequence jumps to line 200.

A simple program to use this instruction is given below. It involves using the input simulator described in Chapter 5.

PROG. 9.3

BBC	Commodore 64	PET
10 ?65122 = 254	10 POKE 56579,254	10 POKE 59459,254
20 ?65120 = 0	20 POKE 56577,0	20 POKE 59471,0
30 IF ?65120 AND 1 THEN 50	30 IF PEEK(56577)AND 1 THEN 50	30 IF PEEK(59471)AND 1 THEN 50
40 GOTO 30	40 GOTO 30	40 GOTO 30
50 ?65120,254	50 POKE 56577,254	50 POKE 59471,254
60 END	60 END	60 END

Line 10 sets all control lines, except P0, as outputs.
Line 20 switches off all the LEDs.
Line 30 tests for an input on P1.
Line 40 provides a loop for testing the state of P1.
Line 50 switches on the LEDs if an input on P1 is received, by using the input simulator switch.

Testing for a bit clear or signal low

An instruction similar to the previous one can be used but here we are testing that the bit is NOT set. The instruction is:

```
              BBC:        Y  IF N AND(NOT(?65120))THEN X
COMMODORE   64:        Y  IF N AND(NOT(PEEK(56577)))THEN X
             PET:        Y  IF N AND(NOT(PEEK(59471)))THEN X
```

As before Y is the line number, X is the next line to be executed if the result is true and N is the pattern of bits providing the mask.

An alternative is to use the previous instruction but, if the bit is set, repeat the instruction again, and again until the bit is cleared as shown by:

```
PET:30 IF PEEK(59471)AND 2 THEN 30
```

As soon as bit 1 goes low, the program jumps out of the loop.

Again the input simulator can be used to practise on these instructions. The next program, program 9.4, tests for the state of control line P1 going low.

PROG. 9.4

BBC	Commodore 64	PET
10 ?65122 = 253	10 POKE 56579,253	10 POKE 59459,253
20 ?65120 = 0	20 POKE 56577,0	20 POKE 59471,0
30 IF ?65120 AND 2 THEN 30	30 IF PEEK(56577)AND 2 THEN 30	30 IF PEEK(59471)AND 2 THEN 30
40 ?65120 = 253	40 POKE 56577,253	40 POKE 59471,253
50 END	50 END	50 END

Line 10 sets seven output lines and one input (P1).
Line 20 switches off all the LEDs.
Line 30 tests to see if P1 is high and if so remains at line 30.
Line 40 switches on the LEDs as soon as the P1 line goes low.

Setting a bit or making a signal high

Many practical applications require the capability to control single lines, to switch on or off devices without affecting the other lines, and in fact without needing to know the states of the other lines. Let us suppose line P0 controls the lights for your swimming pool and P1 controls the automatic garage door and P2 controls the robot that does the ironing, etc., then it is possible to set any of these into action with the following instruction:

```
          BBC:    Y  ?65120 = ?65120 OR N
COMMODORE  64:    Y  POKE 56577,PEEK(56577)OR N
         PET:    Y  POKE 59471,PEEK(59471)OR N
```

N in this program line is the bit or bits (in denary) which are to be set, and hence make the corresponding output lines high. It looks like a clumsy instruction but then it's the results that count.

The LED control circuit can be used to practise the use of this instruction and the following program keeps the LED on P2 illuminated whatever the conditions.

PROG. 9.5

BBC

```
10 INPUT A
20 ?65122 = 255
30 ?65120 = A
40 ?65120 = ?65120 OR 4
50 GOTO 10
```

Commodore 64

```
10 INPUT A
20 POKE 56579,255
30 POKE 56577,A
40 POKE 56577, PEEK(56577) OR 4
50 GOTO 10
```

PET

```
10 INPUT A
20 POKE 59459,255
30 POKE 59471,A
40 POKE 59471, PEEK(59471) OR 4
50 GOTO 10
```

Clearing a bit or making a signal low

A bit was set by using a clumsy looking logical OR in the previous section. A bit can be cleared using a similar instruction but this time using logical AND. For example:

```
            BBC:        Y ?65120 = ?65120 AND N
COMMODORE   64:         Y POKE 56577,PEEK(56577)AND N
            PET:        Y POKE 59471,PEEK(59471)AND N
```

This instruction is a little more awkward than the previous one, in that N must be the value of all the bits (in denary) which are to be left unaffected. Suppose N=191, then bit 6 weighted as 64 in denary would be cleared. This can be shown by assuming the DR is full of 1s before the instruction is executed.

```
        DR   11111111    (255 denary)
         N   10111111    (191 denary)

AND result   10111111
```

An alternative instruction, to clear a bit, incorporating a NOT for the Commodore 64 is:

Y POKE56577, (PEEK (56577) AND NOT M)

The BBC B and PET lines can be changed likewise. Here M is the value of the bit or bits to be cleared. The LED circuit can again be used to test these instructions, as follows.

PROG. 9.6

BBC

```
10 INPUT M
20 ?65122 = 255
30 ?65120 = 255
40 FOR I = 0 TO 1000
50 NEXT I
60 ?65120 = ?65120 AND NOT M
70 GOTO 10
```

Commodore 64

```
10 INPUT M
20 POKE 56579,255
30 POKE 56577,255
40 FOR I = 0 TO 1000
50 NEXT I
60 POKE 56579,(PEEK(56577)AND NOT M)
70 GOTO 10
```

PET

```
10 INPUT M
20 POKE 59459,255
30 POKE 59471,255
40 FOR I = 0 TO 1000
50 NEXT I
60 POKE 59471,(PEEK(59471)AND NOT M)
70 GOTO 10
```

Line 10 asks for a value for M which will determine which LEDs will be switched off later.

Line 20 sets all control lines as outputs.
Lines 40 and 50 are a time delay to allow the operator to see when a change (a bit
      cleared) takes place.
Line 60 clears bits to the value of M.

10

Analogue to digital converters

There are many sensors on the market (discussed in Chapter 5) which can be used to monitor a wide variety of quantities, such as mass, light, temperature, etc. These sensors are analogue but the microcomputer only uses digital signals and so there must be a conversion from analogue to digital signals in order that the micro can 'understand' the incoming signals. Such a conversion is carried out by an electronic circuit known simply as an analogue to digital converter (ADC).

General requirements of a monitoring system

Figure 10.1 shows a block diagram of a monitoring system. The sensor has to be connected into a voltage producing circuit; that is, a circuit whose output voltage changes as the sensor detects changes in the quantity being monitored. In many cases these circuits can be extremely simple (see Fig. 10.11). The analogue voltage produced is now fed into an ADC which produces a meaningful output for the computer.

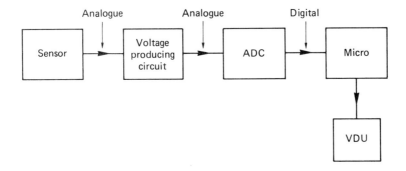

Fig. 10.1 Block diagram of a monitoring system

Principle of the ADC*

An analogue voltage is converted into a corresponding digital signal. A small analogue voltage is converted into a 'small' digital signal—small, not in terms of its voltage, but in terms of its binary or denary value.

* The BBC micro has on-board ADC

The ADC shown in Fig. 10.2 is a theoretical one that can accept inputs in the range of 0 to 5 volt, and these are converted into 8 bit binary signals.

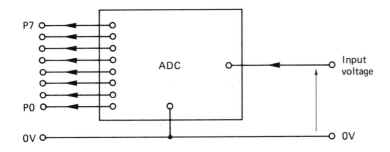

Fig. 10.2 A simple ADC

A 0 V input produces 00000000 (0 denary) output, and a 5V input produces 11111111 (255 denary) output. From these two extremes the relationship between the input and output can be determined. For this particular ADC an increase in input of about 20mV increments the denary output by 1, so its resolution is about 4%. Although this means this device is not extremely accurate, it would be suitable for many applications.

More sophisticated ADC are available than the simple 8 bit one described in this chapter — some have better resolution and can accept negative as well as positive voltages — but in general this type is quite adequate and can be used for a whole range of monitoring applications.

Ready-built ADC circuits are not cheap but are well worth the investment. Certain models offer DAC also on the circuit board, which is extremely useful. For anyone wishing to produce their own, whether for interest or for economy, many electronics and computing magazines provide circuits. A circuit has not been given as this text is not intended as a practical electronics course.

Two basic types of ADC are:
(a)   Step converters
(b)   Successive approximation converters

(a)   Step converters

In order to illustrate the principle let us draw an analogy by considering the use of a balance to find the mass of an object.

The object of unknown mass is placed on the balance pan (Fig. 10.3). The STEP technique to find the mass is to place one mass on the other pan and see if balance is obtained. If balance is not achieved another mass and then another is placed on the pan. This gradual increase in mass eventually achieves balance — or does it? Consider the graph of the procedure shown in Fig. 10.3.

The STEPPING process is obvious, I hope. If the masses being applied were, say 1 gm, then reference to the graph (Fig. 10.4) reveals that eleven masses were not enough to balance the object. The twelfth mass takes us over the objects

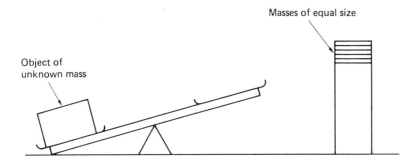

Fig. 10.3 Mass balance

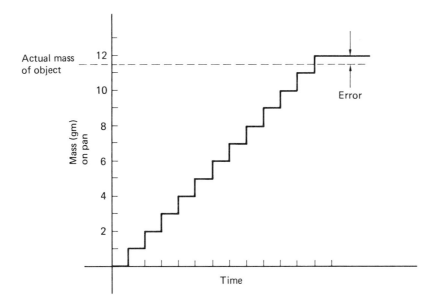

Fig. 10.4 Graph for object mass determination

mass and so balance is not obtained. The approximate mass has been found but the error in the measurement is just under 1 gm.

The stepping ADC operates in a very similar way. It increments the voltage in 20mV steps and compares it with the input analogue voltage. Naturally with the ADC, as with the balance, an error occurs due to the size of step, and the time for conversion increases as the input voltage increases.

(b)   Successive approximation converters

To explain the principle of this type of converter the balance analogy will again be used. This time the mass are not of equal size but are related by a common multiplier of 2 as shown in Fig. 10.5.

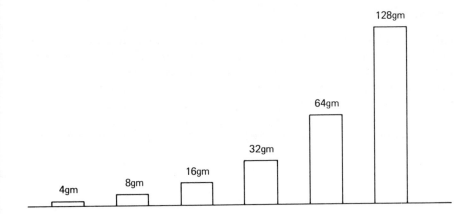

Fig. 10.5 Masses related by multiplying factor of 2

To determine the mass of the object, the largest mass 128 gm is placed on the pan and the outcome determines which mass will be placed on the pan next. This is in a way a trial and error method and again balance can only occur if the object mass exactly equals that of the masses themselves. Hopefully, this analogy gives some understanding of the operation of successive approximation converters. This type of converter is popular because of its speed of conversion and economy.

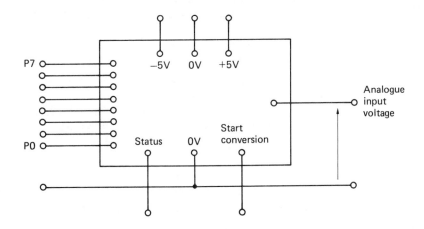

Fig. 10.6 Typical wiring diagram for an ADC

Operating the ADC

A typical arrangement of an ADC in use is given in Fig. 10.6. The ADC circuit itself requires a power supply and a typical supply is shown. Two other lines are shown — convert command and status.

Convert command

Sometimes called **start conversion**. Simply connecting an analogue voltage to the ADC input will not automatically produce a digital output. The ADC needs to be 'told' when to do a conversion — and not just once but every time.

A signal from the user port can be used to start conversion. Let us assume for our ADC that the convert command trigger is **positive edge sensitive**. This means it requires the positive edge of a signal (a change from 0V to 5V) to start the conversion process (Fig. 10.7). A change from 5V to 0V (negative edge) will not produce a conversion.

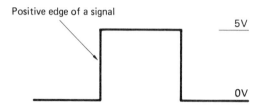

Fig. 10.7 The positive edge of a digital signal

Status

Sometimes called **end of conversion**. Conversions are completed in fractions of a second. In fact, times vary greatly from converter to converter. Some can complete over 100,000 conversions per second. This does not mean that we require this number; we may only require say, five per second.

If the ADC output is read before conversion is complete an incorrect result will be obtained. To allow the correct readings to be secured the ADC indicates the end of conversion. The valid data can now be accepted by the microcomputer and conversion can be reinitiated by the convert command line.

Let us assume with our ADC that the voltage on the status line is 5V during conversion and falls to 0V when conversion is complete and data is valid.

Summing this up for our ADC:

A change from 0 to 1 (0V to 5V) is required to start conversion, and a change from 1 to 0 (5V to 0V) on the status line indicates conversion is complete.

Software for this particular converter is given but if you obtain one with different status and convert command requirements then slight program modifications will naturally be required.

In many ADC the digital representation of the analogue voltage is stored in the output latches of the ADC and this digital information will not be output to the user port unless instructed. With these types the microcomputer instructs the ADC to make the data present at its output latches. This is known as the ENABLE state. The data is received and then the ENABLE signal from the micro reverts back to its original state and the cycle is complete. This is known as the DISABLE state. A new cycle is started with the necessary signal on the convert command line.

Because simplicity and basic understanding is desired with this text the software given here is for the simplest ADC, i.e. those without the ENABLE and DISABLE states, so that the output data can be picked up at any time.

The most satisfactory method of controlling and monitoring an ADC is using the handshake lines, but because this is rather difficult to grasp for the average beginner I have left this until later. The method described now works extremely well indeed and is much simpler than the handshake technique. It has, however, one inherent shortcoming as will soon become obvious.

A look back at Fig. 10.6 reveals that ten lines are necessary for the control and monitoring of the ADC.
These are:

8 inputs to the user port for receiving the digital data
1 input from the status
1 output to start conversion

If the handshake lines are not to be used then at our disposal are the eight lines P0–P7. The two lines with least significance in terms of their weighting are P0 and P1 and these can be connected to the convert command and status respectively. Figure 10.8 is the new wiring diagram. Only a small fraction of the incoming data to the micro has been lost by doing this modification. The maximum input to the micro is now 252, which is only a loss of about 1%.

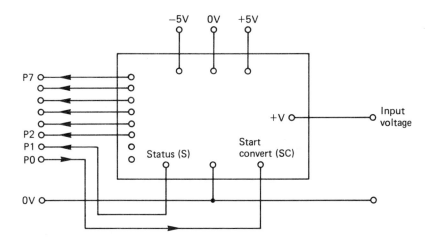

Fig. 10.8 Modified ADC wiring diagram

The basic ADC program

Setting the user port lines is the first requirement. There are to be seven inputs P1–P7 and one output P0 and so the first line of the program is:

```
      BBC                  COMMODORE 64            PET

10 ?65122=1       10 POKE56579,1       10 POKE59459,1
```

A conversion can now be initiated and the positive edge of a signal is required. This is provided by programming in '0' and then a '1' on line P0 as follows:

```
      BBC                  COMMODORE 64            PET

20 ?65120=0       20 POKE56577,0       20 POKE59471,0
30 ?65120=1       30 POKE56577,1       30 POKE59471,1
```

Conversion is now underway and it is necessary to monitor the STATUS line to determine when the conversion process is complete.

```
      BBC                  COMMODORE 64              PET

40 IF ?65120 AND     40 IF PEEK(56577)AND     40 IF PEEK(59471)AND
   2 THEN 40             2 THEN 40                2 THEN 40
```

During the conversion process the status line is 'high' and so line 40 is used to determine when the status goes 'low'. As soon as this occurs the valid data is available and its denary value can be printed on the screen and the cycle started again by:

```
      BBC                  COMMODORE 64              PET

50 PRINT ?65120      50 PRINT PEEK(56577)     50 PRINT PEEK(59471)
60 GOTO 20           60 GOTO 20               60 GOTO 20
```

To reduce the number of conversions per second displayed on the screen a time delay could be introduced between lines 50 and 60.

```
      BBC                  COMMODORE 64                PET

52 FOR I=0 TO 200    52 FOR I=0 TO 200       52 FOR I=0 TO 200
55 NEXT I            55 NEXT I               55 NEXT I
```

This basic program is suitable for many ADCs and can be used whatever quantity is being monitored. How the data is manipulated and displayed once the microcomputer has received it, however, depends on the programmer.

Measurement of calculator battery voltage

The easiest quantity to measure using an ADC is voltage because no transducer/sensor is required. The voltage of a calculator battery can be determined by connecting the battery directly to the input of the ADC (see Fig. 10.9). Running the basic ADC program now should produce values on the screen as follows:

75
75
75
75
75
75
.
.
.

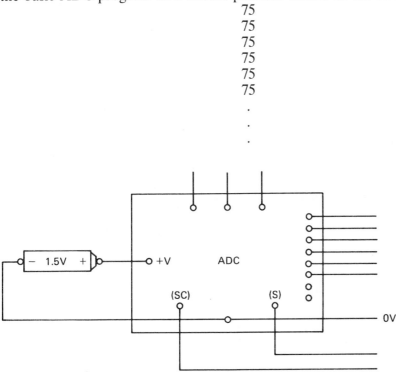

Fig. 10.9 Measurement of voltage

Why is this?
The answer is quite simply that each denary unit represents approximately 20mV and so 1.5 volt/0.02 volt = 75. Connecting a $4\frac{1}{2}$V battery to the ADC should produce readings of about 225.

To program the microcomputer to display actual voltage values means manipulating the data as shown in the complete program below.

Monitoring light intensity

Light intensity can be monitored by the microcomputer by utilising the properties of an LDR (light dependent resistor). Such components are cheap and reliable and are extremely easy to use.

The resistance of an LDR varies considerably with the degree of illumination. A typical one (ORP 12) has a resistance of about 10MΩ in darkness and this falls to just over 100Ω in good light. A simple investigation into the properties of an LDR is to connect it to an ohmmeter and then vary the light falling on it, as shown in Fig. 10.10.

PROG. 10.1 Measurement of voltage

BBC

```
10 ?65122=1
20 ?65120=0
30 ?65120=1
40 IF ?65120 AND 2 THEN 40
50 A=?65120
60 V=A*0.02
70 PRINT V;"VOLTS"
80 GOTO 20
```

COMMODORE 64

```
10 POKE56579,1
20 POKE56577,0
30 POKE56577,1
40 IF PEEK(56577)AND 2 THEN 40
50 A=PEEK(56577)
60 V=A*0.02
70 PRINT V;"VOLTS"
80 GOTO 20
```

PET

```
10 POKE59459,1
20 POKE59471,0
30 POKE59471,1
40 IF PEEK(59471)AND 2 THEN 40
50 A=PEEK(59471)
60 V=A*0.02
70 PRINT V;"VOLTS"
80 GOTO 20
```

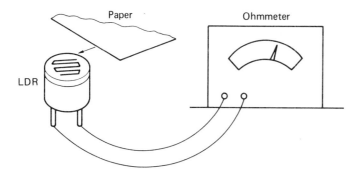

Fig. 10.10 Investigation into the properties of an LDR

Voltage production

An LDR does not produce an analogue voltage by itself and so it must be incorporated in a 'voltage producing circuit'. The necessary circuit is surprisingly simple as can be seen in Fig. 10.11.

To understand its operation it is essential to realise that the circuit consists basically of two resistors in series. The supply is a constant 5V and so any variation in the resistance of the LDR changes the voltage across each component. This voltage variation can be picked off either the LDR or the resistor and can then be fed into the ADC.

The value of the accompanying resistor depends on the LDR used. Basically what is required is a circuit that produces as large a variation in voltage between dark and light conditions as possible. Naturally the ideal conditions are that

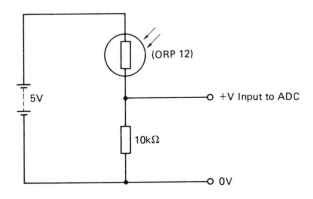

Fig. 10.11 Voltage producing circuit for an LDR

when dark the voltage across the LDR should approach 5V and when light the voltage across the resistor should approach 5V. If these conditions are not met then the system will operate but will lack sensitivity.

The voltage drops across the two components can be determined for dark and light conditions as follows:

DARK

$$\begin{array}{lll} \text{LDR} & & 1\text{M}\Omega \\ \text{Rx} & = & 10\text{k}\Omega \end{array}$$

The ratio of resistances is $\dfrac{10\text{M}}{10\text{K}} = 1000$ to 1.

This means that the voltage across the LDR is 1000 times that across the resistor and the voltages will be about 4.995V and 0.005V respectively.

LIGHT

$$\begin{array}{lll} \text{LDR} & & 100\Omega \\ \text{Rx} & = & 10\text{k}\Omega \end{array}$$

The ratio of resistances in light is 100 to 1 in the opposite direction. The voltages across the LDR and Rx will be approximately 0.05V and 4.95V respectively, and good sensitivity has therefore been achieved.

With the light sensitive circuit connected to the user port via the ADC the basic ADC program displays numbers in the range 0 to 252 on the screen as light intensity varies. Program 10.2 is a modification and displays light and dark states.

PROG. 10.2 Monitoring light intensity

BBC

```
10 REM MONITORING LIGHT
20 ?65122=1
30 ?65120=0
40 ?65120=1
50 IF ?65120 AND 2 THEN 50
60 A=?65120
70 IF A<50 THEN PRINT"DARK"
80 IF A>120 THEN PRINT"LIGHT"
90 FOR I=0 TO 200
100 NEXT I
110 GOTO 30
```

COMMODORE 64

```
10 REM MONITORING LIGHT
20 POKE59459,1
30 POKE59471,0
40 POKE59471,1
50 IF PEEK(59471)AND 2 THEN 50
60 A=PEEK(59471)
70 IF A<50 THEN PRINT"DARK"
80 IF A>120 THEN PRINT"LIGHT"
90 FOR I=0 TO 200
100 NEXT I
110 GOTO 30
```

PET

```
10 REM MONITORING LIGHT
20 POKE59459,1
30 POKE59471,0
40 POKE59471,1
50 IF PEEK(59471)AND 2 THEN 50
60 A=PEEK(59471)
70 IF A<50 THEN PRINT"DARK"
80 IF A>120 THEN PRINT"LIGHT"
90 FOR I=0 TO 200
100 NEXT I
110 GOTO 30
```

This simple light monitoring system can be transformed into an object counting system. Figure 10.12 shows a practical situation where objects on a conveyor belt are being counted.

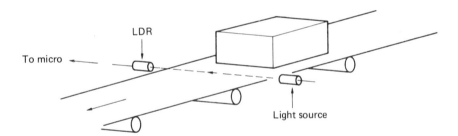

Fig. 10.12 Object counter

The software naturally should increment the count by one each time the light intensity on the LDR falls to a specified level.

Calibration of the measuring system

Just as measuring instruments require calibration against some standard, so a microcomputing measuring system employing some sensor also requires calibration before it can be used.

Unfortunately, the characteristics of many sensors are non-linear so calibration is not simple. Suppose, for example, we were measuring temperature using a thermistor as the sensor. We should now run the ADC system and at the same time record true temperatures with some instrument that gives accurate temperature readings. A calibration curve can then be constructed as shown in Fig. 10.13.

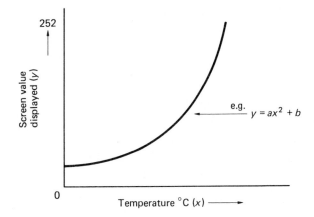

Fig. 10.13 A calibration curve

One method of calibrating the microcomputer system now is to find an equation for the curve—or one that is a very close fit, e.g. $y = ax^2 + b$. This equation can now be introduced into the software so that temperature is cal culated from the value of the input to the micro.

If this method proves unsatisfactory an alternative is to store numerous values (taken from the calibration curve) in memory and use the software to select the correct or nearest correct temperature.

This particular problem is discussed in Chapter 12.

11

Handshake

Handshaking is a technique which allows the efficient transfer of data between the computer and an external device.

Most commonly used peripherals respond, and act, far too slowly for the computer and so some form of synchronisation of the two must be effected. This synchronisation is achieved by using control or handshake lines (in addition to the normal data lines) between the two items. Signals on these lines may indicate such things as 'data ready', 'data received', etc.

A typical dialogue between the computer and a peripheral might be:

Computer   — 'Are you ready to receive data?'
Peripheral   — 'Yes'/'No'

This is illustrated in Fig. 11.1.

The dialogue, albeit less refined than that above, ensures that data is not lost due to the different operating speeds of the computer and peripheral.

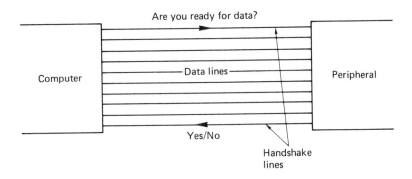

Fig. 11.1 Handshake dialogue

Handshake lines

Many microcomputers are equipped with two handshake lines at the user port in addition to the eight input/output lines. The handshake lines for the Commodore 64, BBC and PET microcomputers are as follows:

88

	C1	C2
COMM 64	Flag 1	PA2
BBC	CB1	CB2
PET	CA1	CB2

Unfortunately, there are many different ways in which these lines can be used. For example, CB2 on the BBC can be configured to perform in eight different modes. Because of this only the most useful, at this level of control, will be described.

C1 for each of the three micros listed above, can only be used as an input line. This control line is edge sensitive in that it will only respond to either a change from logic 0 (0V) to logic 1 (5V) or vice versa. It can be programmed to respond to one edge or the other, but it cannot be made to respond to both simultaneously.

The C2 lines in contrast can be set as an output or as an input, but here they will be used as output lines.

In Chapter 10 it was suggested that the ADC could be controlled using handshaking and so here I propose to explain the principles of handshaking by relating it directly to the control of the ADC. Because C1 can be used as an input line it is ideally suitable for detecting the **end of conversion** on the status line. Likewise, because C2 can be set to act as an output it can be used to **start conversion**. Figure 11.2 shows the ADC ready for control through handshaking.

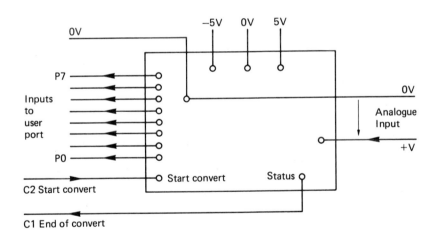

Fig. 11.2 Handshake control of the ADC

Control and interrupt registers

Control of the handshake lines (except for the Commodore 64) is primarily through the use of two registers — the Peripheral Control Register (PCR) and the Interrupt Flag Register (IFR).

The PCR sets the mode of operation of the handshake lines. Particular bits set or cleared in this register determine whether C2 is to be output or input and whether C1 is positive or negative edge sensitive. The programmer naturally sets or clears these bits to achieve his particular objective.

Once C1 has been set and the particular signal edge is detected a bit (or flag) is set in the IFR. In the case of controlling the ADC it is necessary to inspect the IFR continually to detect the setting of the relevant bit to determine when conversion is complete.

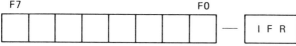

Particular bits set in this register control
the mode of C1 and C2

D7   D0

| | | | | | | | | — | P C R |

A bit is set in the IFR on detection of a signal edge

F7   F0

| | | | | | | | | — | I F R |

Fig. 11.3 IFR and PCR

BBC handshaking

The PCR for the BBC is located at 65132 in memory and the IFR is located at 65133.

CB1 line

CB1 is an edge sensitive input line and all that the programmer has to do is to select which edge of a signal the line will be sensitive to.

To set CB1 to be positive edge sensitive (a change from 0V to 5V) the 16 weighted (D4) bit in the PCR must be set, as follows:

?65132 = ?65132   OR   16

The same bit must be cleared to make CB1 negative edge sensitive and this is achieved by:

?65132 = ?65132   AND   239

Figure 11.4 illustrates this in diagram form.

When CB1 is set, only the programmed edge will be sensed and on sensing the edge of a signal the 16 weighted (F4) bit in the IFR will be set. This is illustrated in Fig. 11.5.

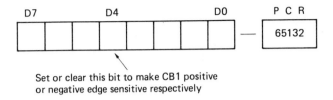

Set or clear this bit to make CB1 positive
or negative edge sensitive respectively

Fig. 11.4 PCR

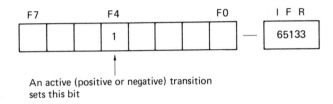

An active (positive or negative) transition
sets this bit

Fig. 11.5 IFR

To detect when the F1 bit is set we must monitor the IFR using the following loop.

```
80 IF ?65133 AND 16 THEN 100
90 GOTO 80
100
```

The program sequence remains in the loop until the F4 bit is set. If the ADC is being controlled then the setting of this bit indicates the end of conversion.

This bit, once set by an active signal edge, will remain set and so it must be cleared so that the next active transition (end of conversion) can be indicated. Clearing this presents no real problems. Simply writing to, or reading from, the DR located at 65120 automatically clears the bit. A write instruction such as ?65120 = 0 will achieve this.

CB1 Summary

Let us summarise the control sequence for the CB1 line.

(a) Set or clear the D4 bit in the PCR to make CA1 positive or negative edge sensitive.
(b) Monitor the IFR to detect when the F4 bit is set by detection of a signal edge, which could be say, to indicate end of conversion for an ADC.
(c) Clear the F4 bit in the IFR by reading from or writing to the DR at 65120.

CB2 line

The BBC CB2 can be set as an input or output line. Bits 5, 6 and 7 in the PCR control the CB2 mode of operation (Fig. 11.6) and there are eight modes in all.

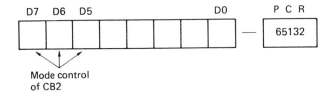

Mode control
of CB2

Fig. 11.6

The PCR bits 5, 6 and 7 have the effects shown in Table 11.1.

Table 11.1

Bits D7 D6 D5	Denary equivalent	Effect
0 0 0	0	Input mode — negative edge sensitive IFR bit 4 cleared by a read or write to 65120.
0 0 1	32	Input mode — as 0,0,0 but bit 4 is not cleared by a read or write.
0 1 0	64	Input mode — as for 0,0,0 but positive edge sensitive.
0 1 1	96	Input mode — as for 0,0,1 but positive edge sensitive.
1 0 0	128	Output mode — CB2 is set high by an active transition on the CA1 input. Reset by reading or writing to 65120 register.
1* 0 1	160	Output mode — CB2 goes low for one cycle of the 1MHz clock, following a read or write of the 65120 register.
1 1 0	192	Output mode — CB2 always low.
1 1 1	224	Output mode — CB2 always high.

If the CB2 line is to be used to start conversion for an ADC then the most convenient mode is the one marked thus *. The line goes low for one clock cycle and so there will be a change from logic 1 to logic 0 and back to logic 1. This is enough to initiate a conversion. CB2 can be set in this mode using the instruction:

$$?65132 = ?65132 \quad OR \quad 160$$

To produce the pulse to start conversion it is necessary to read from or write to the DR at 65120. Using

$$?65120 = 0$$

would achieve this.

The complete ADC control program with handshaking is given in program 11.1.

PROG. 11.1 Handshake control of ADC (BBC)

```
10 ?65122 = 0
20 ?65132 = ?65132 OR 160
30 ?65132 = ?65132 AND 239
40 ?65120 = 0
50 IF ?65133 AND 16 THEN 70
60 GOTO 50
70 PRINT ?65120
80 GOTO 40
```

Line 10 sets all the data lines P0/P7 as inputs.
Line 20 sets CB2 as an output line (mode 1,0,1).
Line 30 sets CB1 as a negative edge sensitive line.
Line 40 sends a pulse to start conversion.
Lines 50 and 60 hold the program in a loop until the F4 bit IFR is set indicating
      end of conversion.
Line 70 prints the value (in denary) found in the DR and simultaneously clears
      the F4 bit.
Line 80 starts the conversion sequence again.

Commodore 64 handshaking

Flag 1 line

This line can only be used as an input and the line itself is negative edge sensitive. It only responds to changes from logic 1 (5V) to logic 0 (0V).

On detecting a negative transition the F4 bit in the IFR (located at 56589 in memory) is set. To detect the setting of this bit we must monitor the IFR using the following loop:

```
80 IF PEEK ( 56589 ) AND 16 THEN 100
90 GOTO 80
100
```

Once set this bit remains set and must be cleared so that it can be used to indicate the next active transition (end of conversion). A simple write to or read from the DR at 56577 achieves this automatically.

PA2 line

There is no CB2 line at the Commodore 64 user port but a PA2 line brought out from PORT A* is available and can be used in a similar way to the CB2 line on the BBCB.

Control of the PA2 line is not through the PCR as is the control of the CB2 lines. Controlling the PA2 line is just like controlling any of the P0/P7

---

* PORT A and PORT B are from an interface chip within the microcomputer. The Commodore 64 uses port B as the user port only, and PA2 from port A is available.

input/output lines. This is done by setting the line as either an output or an input by placing a '1' or '0' in the DDR for port A. Signals can then be sent (or received) by using the DR for port A.

The DDR for port A is located at 56578 and the DR is a 56576.

PA2 is controlled by the 4 weighted (P2) bit and so to set this as an output line the following instruction can be used:

```
POKE 56578,( PEEK (56578) OR 4)
```

Here the 4 weighted bit is set without affecting the other lines in port A. One might have thought POKE 56578,4 would suffice. It would certainly set the bit we require, but what else would it upset?

PA2 can be set as an input by clearing the same bit as follows:

```
POKE 56578,(PEEK  (56578) AND 251)
```

If the PA2 line is to be used to start conversion for an ADC then it must be set as an output using the first instruction above. To start conversion requires a change from logic 0 to logic 1 and this can be achieved by:

```
POKE 56576,( PEEK (56576) AND 251) : REM PA2 low
POKE 56576,( PEEK (56576) OR 4)    : REM PA2 high
```

The complete ADC control with handshaking is given in program 11.2.

PROG. 11.2

```
10 POKE 56579,0
20 POKE 56578 (PEEK (56578) OR 4)
30 POKE 56576 (PEEK (56576) AND 251)
40 POKE 56576 (PEEK (56576) OR 4)
50 IF PEEK (56589) AND 16 THEN 70
60 GOTO 50
70 PRINT PEEK (56577)
80 GOTO 20
```

Line 10 sets all the input/output lines as inputs.
Line 20 sets PA2 as an output line.
Lines 30 and 40 produce an output on PA2 to start conversion.
Line 50 tests to see if the 16 weighted bit in the IFR has been set, i.e. it tests to see if conversion is complete.
Line 60 maintains the testing loop.
Line 70 prints the value found in the DR and simultaneously clears the interrupt flag.

PET handshaking

The PCR for the PET is located at 59468 in memory and the IFR is located at 59469.

CA1 line

It has been mentioned already that this line can only be used as an input. To set CA1 to be positive edge sensitive (a change from 0V to 5V) the 1 weighted (D0) bit in the PCR must be set. This can be achieved by:

```
POKE 59468, (PEEK (59468) OR 1)
```

To make a negative edge sensitive line out of CA1 the D0 bit must be cleared and this can be done by:

```
POKE 59468, (PEEK (59468) AND 254)
```

Figure 11.7 illustrates this diagramatically.

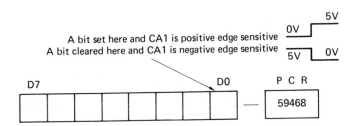

Fig. 11.7

Once the CA1 line mode has been fixed only the 'programmed edge' will be sensed, and as soon as the active positive (or negative signal) edge is detected the 2 weighted (F1) bit in the IFR will be set. This is illustrated in Fig. 11.8.

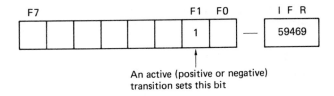

Fig. 11.8 IFR

To detect when the F1 bit is set we must monitor the IFR using the following loop;

```
80 IF PEEK (59469) AND 2 THEN 100
90 GOTO 80
100
```

The program sequence is held at lines 80 and 90 until the F1 bit is detected. Alternatively, a WAIT statement can be used and the following program line achieves the same result.

```
80 WAIT 59469,2,0
```

If the ADC is being controlled then the setting of this bit indicates the end of conversion.

Unfortunately, once the F1 bit has been set it will remain set and so it must now be cleared ready for indicating the next active signal/transition (end of conversion). The most convenient way of doing this is to clear the bit auto matically within the sequence as described below.

Up until now we have observed incoming data (signals) to the user port by examining the DR at 59471. There is another DR where this data is stored and this is located at 59457. We can therefore use this register instead of 59471 and the advantage it offers is that when you write (POKE) to it or read (PEEK) from it you also clear the F1 bit in the IFR. In other words use:

> DR 59471 — for general control
> DR 59457 — for handshake control

CA1 summary

All this may seem complicated so let us summarise the use of CA1.

(a) Set or clear the PCR D0 bit to make CA1 positive or negative edge sensitive.
(b) Monitor the IFR to observe when the F1 bit is set by detection of a signal edge — for example this may be the end of conversion of an ADC.
(c) Clear the F1 bit in the IFR by simply reading or writing to the DR 59457.

CB2 line

As was previously stated this signal line can be programmed as an input or as an output line, but because the other handshake line can only be used as an input line CB2 is generally used in output mode. The mode of operation of this line is also controlled by the PCR at 59468.

The 128 weighted (D7) bit and the 64 weighted (D6) bit must be set in the PCR for CB2 to be in output mode (see Fig. 11.9)

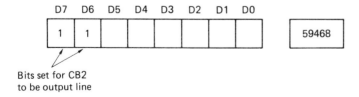

Bits set for CB2
to be output line

Fig. 11.9 Peripheral Control Register

These bits can be set by using the instruction:

POKE 59468, (PEEK (59468)   OR   128   OR   64)
more simply
OR   192

Once CB2 has been set as an output signals can be put on this line by setting or clearing the D5 bit in the PCR. If D5 is set then the CB2 line will be at logic 1 (5V). If D5 is cleared then CB2 will be at logic 0 (0V).

Setting CB2 at logic 1 (5V)   POKE 59468, (PEEK (59468) OR 32)

Setting CB2 at logic 0 (0V)   POKE 59468, (PEEK (59468) AND NOT 32)

or alternatively   POKE 59468, (PEEK (59468) AND 223)

The ADC described in Chapter 10 required a change from 0V to 5V on the 'convert command' line to initiate a conversion. CB2 can be used to start conversion by first programming it to perform as an output line, then setting it at 0V (logic 0), then setting it at 5V (logic 1). To start conversion requires the following program lines:

```
10 POKE 59468, (PEEK (59468) OR 192)
20 REM CB2 set as an output
30 POKE 59468, (PEEK (59468) AND 223)
40 REM CB2 set at logic 0
50 POKE 59468, (PEEK (59468) OR 32)
60 REM CB2 set at logic 1
```

The complete ADC program under handshake control is given in program 11.3.

PROG. 11.3 ADC with Handshaking (PET)

```
10 POKE 59459,0
20 POKE 59468 (PEEK (59468) OR 128 OR 64)
30 POKE 59468 (PEEK (59468) AND 254)
40 POKE 59468 (PEEK (59468) AND 223)
50 POKE 59468 (PEEK (59468) OR 32)
60 IF PEEK (59469) AND 2 THEN 80
70 GOTO 60
80 PRINT PEEK (59459)
90 GOTO 20
```

Line 10 sets all the data lines P0/P7 as inputs.
Line 20 sets CB2 as an output line.
Line 30 sets CA1 as a negative edge sensitive line.
Line 40 sets CB2 at logic 0   start conversion.
Line 50 sets CB2 at logic 1

Lines 60 and 70 hold the program in a loop until the F1 bit in the IFR is set
    indicating end of conversion.

Line 80 reads the contents of the DR and automatically clears the F1 bit.

Line 90 starts the sequence again.

12

Miscellaneous projects

(a) Traffic light simulator

Probably the easiest circuit building project that even the non-electronics enthusiasts can undertake is the traffic light simulator. It combines a simple control circuit with a useful introduction to control software, in both BASIC and assembly language.

The circuit design for the simulator (Fig. 12.1) is similar to the LED control circuit described in Chapter 4. Here, however, sixteen LEDs are required to be switched by eight transistors. Naturally red, yellow and green LEDs are required. A1 kΩ resistor can be connected between the computer and the base of the transistors if desired.

In order that some semblance to the real situation is obtained the circuit board arrangement could be as illustrated in Fig. 12.2, with the NW/ES lanes indicated. Table 12.1 provides the logic sequence along with the numbers that must be 'POKED' to the user port to achieve this sequence. Green filter lights have not been included at this stage. The corresponding program follows.

Table 12.1 Traffic light logic sequence

E/W	N/S	E/W				N/S				
		128 G Filter	64 G	32 A	16 R	8 G Filter	4 G	2 A	1 R	Value at DR
Red	Red/Amber	0	0	0	1	0	0	1	1	19
Red	Green	0	0	0	1	0	1	0	0	20
Red	Amber	0	0	0	1	0	0	1	0	18
Red/Amber	Red	0	0	1	1	0	0	0	1	49
Green	Red	0	1	0	0	0	0	0	1	65
Amber	Red	0	0	1	0	0	0	0	1	33

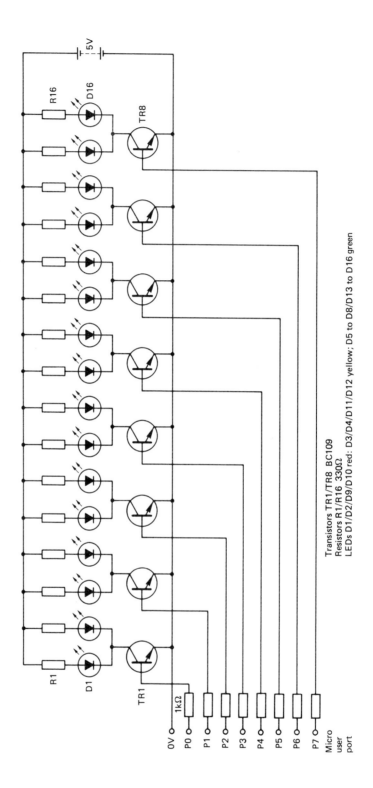

Fig. 12.1 Traffic light simulator circuit

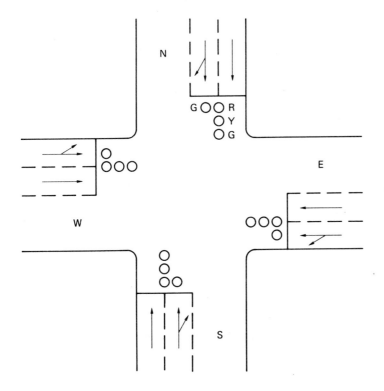

Fig. 12.2 Plan view of simulator

PROG. 12.1

BBC

```
 10 REM TRAFFIC LIGHTS
 20 ?65122=255
 30 ?65120=19
 40 GOSUB 500
 50 ?65120=20
 60 GOSUB 500
 70 ?65120=18
 80 GOSUB 500
 90 ?65120=49
100 GOSUB 500
110 ?65120=65
120 GOSUB 500
130 ?65120=33
140 GOSUB 500
150 GOTO 30
500 REM DELAY
510 FOR I=0 TO 1000
520 NEXT I
530 RETURN
```

COMMODORE 64

```
10 REM TRAFFIC LIGHTS
20 POKE56579,255
30 POKE56577,19
40 GOSUB 500
50 POKE56577,20
60 GOSUB 500
70 POKE56577,18
80 GOSUB 500
90 POKE56577,49
100 GOSUB 500
110 POKE56577,65
120 GOSUB 500
130 POKE56577,33
140 GOSUB 500
150 GOTO 30
500 REM DELAY
510 FOR I=0 TO 1000
520 NEXT I
530 RETURN
```

PET

```
10 REM TRAFFIC LIGHTS
20 POKE59459,255
30 POKE59471,19
40 GOSUB 500
50 POKE59471,20
60 GOSUB 500
70 POKE59471,18
80 GOSUB 500
90 POKE 59471,49
100 GOSUB 500
110 POKE 59471,65
120 GOSUB 500
130 POKE 59471,33
140 GOSUB 500
150 GOTO 30
500 REM DELAY
510 FOR I=0 TO 1000
520 NEXT I
530 RETURN
```

Line 10 sets all control lines as outputs.
Lines 30, 50, 70 etc., provide the required traffic light sequence.
Lines 40, 60, 80 etc., bring in the time delay housed in the subroutine.

This software is very basic and leaves much room for change and improvement. Only one time delay is used, and in practice at least two would be required, as the amber lights are on for only a short time in relation to the others. Further software could include the use of a key to simulate traffic flow and produce a response on the LED traffic board.

(b) Pedestrian crossing simulator

Although the circuit (Fig. 12.3) for this simulator uses transistor switching, and is very similar to the traffic light simulator circuit, much more interesting software can be introduced. Besides controlling LEDs, sound must be generated and an input signal used to effect the start of the crossing cycle.

The bleeping sound can be produced in two ways. It can either be obtained by switching in a loudspeaker and the necessary accompanying circuitry, or it could be generated by the microcomputer itself. The second option has been selected for economy and convenience.

An input signal must be derived (a button pressed by the pedestrian) and this can also be done in two ways. A push-to-make switch on the circuit board could switch in a 5V supply to produce an input signal to the user port via the input circuit described in Chapter 3. Alternatively a button on the microcomputer keyboard can be used. The latter has been selected here. Figure 12.4 shows the LED layout for the simulator.

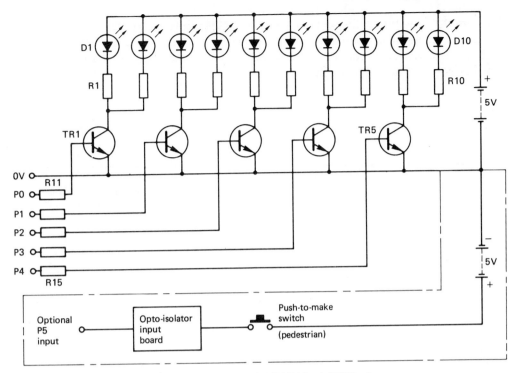

LEDs D1/D2/D5/D6 green; D3/D4/D9/D10 red; D7/D8 yellow
Resistors R1/R10 330Ω
Transistors TR1/TR5 BC109
Resistors R11/R15 1kΩ

Fig. 12.3 Pedestrian crossing simulator circuit

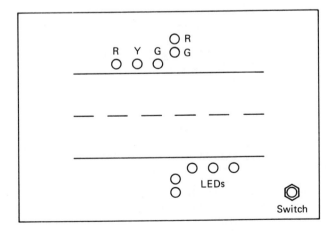

Fig. 12.4 Plan view of pedestrian simulator

Program 12.2 assumes the use of key 'S' on the microcomputer keyboard to initiate the crossing sequence. No sound is generated by the micro; this can be a further exercise for the reader. First, however, we must state the actual sequence and this is given in Table 12.2.

Table 12.2 Pedestrian crossing sequence

USER PORT

Traffic			Pedestrian		User port output value	Button pressed to initiate cycle	Sound generated
Red (16)P4	Yellow (8)P3	Green (4)P2	Red (2)P1	Green (1)P0			
0	0	1	1	0	6	NO	NO
0	1	0	1	0	10	YES	NO
1	0	0	0	1	17	NO	YES
0	1 (flash)	0	0	1 (flash)	9	NO	NO

PROG. 12.2

BBC

```
 10 ?65122=255
 20 REM**GREEN AND RED**
 30 ?65120=6
 40 PRINT "PRESS 'S' TO CROSS"
 50 A$=GET$
 60 IF A$="S" THEN 90
 70 REM S STARTS CYCLE
 80 GOTO 50
 90 REM**AMBER AND RED**
100 ?65120=10
110 GOSUB 500
120 REM**RED AND GREEN**
130 ?65120=17
140 GOSUB 500
150 REM**FLASHING LEDS**
160 X=0
170 ?65120=9
180 GOSUB 600
190 ?65120=0
200 GOSUB 600
210 X=X+1
220 IF X<10 GOTO 170
230 GOTO 20
500 FOR I=0 TO 1500
510 NEXT I
520 RETURN
600 FOR J=0 TO 150
610 NEXT J
620 RETURN
```

COMMODORE 64

```
 10 POKE56579,255
 20 REM**GREEN AND RED**
 30 POKE56577,6
 40 PRINTTAB(5)"PRESS 'S' TO CROSS"
 50 GET A$: IF A$=" " THEN 50
 60 REM 'S' STARTS CYCLE
 70 IF A$="S" THEN 90
 80 GOTO 50
 90 REM**AMBER AND RED**
100 POKE56577,10
110 GOSUB 500
120 REM**RED AND GREEN**
130 POKE56577,17
140 GOSUB 500
150 REM**FLASHING LEDS**
160 X=0
170 POKE56577,9
180 GOSUB 600
190 POKE56577,0
200 GOSUB 600
210 X=X+1
220 IF X<10 GOTO 170
230 GOTO 20
500 FOR I=0 TO 1500
510 NEXT I
520 RETURN
600 FOR J=0 TO 150
610 NEXT J
620 RETURN
```

(c)   Combination lock with an alarm

This is simply a programming exercise assuming you have at your disposal the LED control board and either a stepper motor or a d.c. motor.

Basically the system simulates a microcomputer controlled safe combination lock incorporating an alarm device. The software can be such that the operator (or thief) keys in say, four numbers and if the correct sequence is keyed in the motor rotates to open the safe. Naturally, if the incorrect sequence is entered the alarm sounds and the LEDs flash continuously.

The circuit is given (Fig. 12.5) and here a stepper motor is being used.

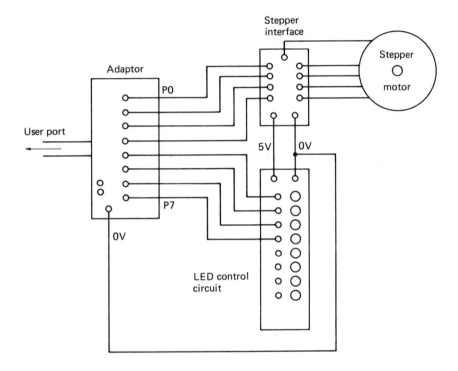

Fig. 12.5 Safe lock simulation circuit

The following software provides a lock combination of 9, 2, 7, 4. No sound is generated and this can be a further exercise for the programmer.

(d)   Stepper motor control using switches

It is likely that stepper motors will be used to drive some item rather than be used simply as a programming exercise piece. If this is the case then control could be by the software alone, by a combination of software and keyboard control or by a combination of software and input signals derived when switches are closed. The latter can be done by using switches on the input simulator described in Chapter 5 or alternatively a suitable supply can be switched in to provide input signals.

PROG. 12.3

BBC

```
 10 REM COMBINATION LOCK
 20 ?65122=255
 30 INPUT "INPUT 1ST NUMBER";A
 40 INPUT "INPUT 2ND NUMBER";B
 50 INPUT "INPUT 3RD NUMBER";C
 60 INPUT "INPUT 4TH NUMBER";D
 70 IF A=9 THEN 90
 80 GOTO 300
 90 IF B=2 THEN 110
100 GOTO 300
110 IF C=7 THEN 130
120 GOTO 300
130 IF D=4 THEN 150
140 GOTO 300
150 REM*OPEN SAFE*
155 S=0
160 ?65120=3
170 GOSUB 500
180 ?65120=6
190 GOSUB 500
200 ?65120=12
210 GOSUB 500
220 ?65120=9
230 GOSUB 500
240 S=S+1
250 IF S=20 THEN END
260 GOTO 160
300 REM*ALARM*
310 ?65120=240
320 FOR J=0 TO 50:NEXT J
330 ?65120=0
340 FOR L=0 TO 50:NEXT L
350 GOTO 300
500 REM*MOTOR SPEED CONTROL*
510 FOR I=1 TO 50
520 NEXT I
530 RETURN
```

COMMODORE 64

```
 10 REM COMBINATION LOCK
 20 POKE56579,255
 30 INPUT "INPUT 1ST NUMBER";A
 40 INPUT "INPUT 2ND NUMBER";B
 50 INPUT "INPUT 3RD NUMBER";C
 60 INPUT "INPUT 4TH NUMBER";D
 70 IF A=9 THEN 90
 80 GOTO 300
 90 IF B=2 THEN 110
100 GOTO 300
110 IF C=7 THEN 130
120 GOTO 300
130 IF D=4 THEN 150
140 GOTO 300
150 REM*OPEN SAFE*
155 S=0
160 POKE56577,3
170 GOSUB 500
180 POKE56577,6
190 GOSUB 500
200 POKE56577,12
210 GOSUB 500
220 POKE56577,9
230 GOSUB 500
240 S=S+1
250 IF S=20 THEN END
260 GOTO 160
300 REM*ALARM*
310 POKE56577,240
320 FOR J=0 TO 50:NEXT J
330 POKE56577,0
340 FOR L=0 TO 50:NEXT L
350 GOTO 300
500 REM*MOTOR SPEED CONTROL*
510 FOR I=1 TO 50
520 NEXT I
530 RETURN
```

If you are to produce the input simulator then four of the switches could be used to control the motor, one switch each for stop, start, forward and reverse. Figure 12.6 shows the connections for the motor and input simulator.

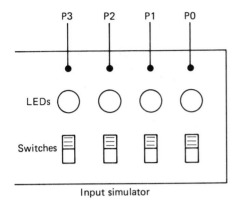 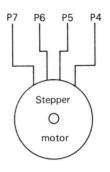

Input simulator

Fig. 12.6

PROG. 12.4

This controls the motor with switches on P0 to P3.

PO = start (1)
P1 = stop (2)
P2 = forward (4)
P3 = reverse (8)

BBC

```
 10 ?65120=240
 20 IF ?65120 AND 1 THEN 200
 30 IF ?65120 AND 2 THEN 100
 40 GOTO 20
100 REM STOP MOTOR
110 ?65120=0
120 GOTO 20
200 REM START MOTOR FORWARD
210 ?65120=192
220 GOSUB 600
230 ?65120=96
240 GOSUB 600
250 ?65120=48
260 GOSUB 600
270 ?65120=144
280 GOSUB 600
290 IF ?65120 AND 2 THEN 100
300 IF ?65120 AND 8 THEN 400
310 GOTO 200
400 REM MOTOR REVERSE
410 ?65120=144
420 GOSUB 600
430 ?65120=48
440 GOSUB 600
450 ?65120=96
```

COMMODORE 64

```
 10 POKE56579,240
 20 IF PEEK(56577)AND 1 THEN 200
 30 IF PEEK(56577)AND 2 THEN 100
 40 GOTO 20
100 REM STOP MOTOR
110 POKE56577,0
120 GOTO 20
200 REM START MOTOR FORWARD
210 POKE56577,192
220 GOSUB 600
230 POKE56577,96
240 GOSUB 600
250 POKE56577,48
260 GOSUB 600
270 POKE56577,144
280 GOSUB 600
290 IF PEEK(56577)AND 2 THEN 100
300 IF PEEK(56577)AND 8 THEN 400
310 GOTO 200
400 REM MOTOR REVERSE
410 POKE56577,144
420 GOSUB 600
430 POKE56577,48
440 GOSUB 600
450 POKE56577,96
```

```
460 GOSUB 600                    460 GOSUB 600
470 ?65120=192                   470 POKE56577,192
480 GOSUB 600                    480 GOSUB 600
490 IF ?65120 AND 2 THEN 100     490 IF PEEK(56577)AND 2 THEN 100
500 IF ?65120 AND 4 THEN 200     500 IF PEEK(56577)AND 4 THEN 200
510 GOTO 400                     510 GOTO 400
600 REM DELAY                    600 REM DELAY
610 FOR I=0 TO 500               610 FOR I=0 TO 500
620 NEXT I                       620 NEXT I
630 RETURN                       630 RETURN
```

Measurement of temperature

A temperature sensor known as a thermistor was briefly described in Chapter 5, and by using this type of sensor in conjunction with an ADC, the microcomputer is able to measure temperature.

Thermistors simply vary in resistance as temperature varies and do not produce an output current or voltage. The thermistor must therefore be connected into a suitable voltage producing circuit, and the simplest is a potential dividing arrangement as shown in Fig. 12.7. A typical thermistor is shown with a suitable accompanying resistor, and the principle of the circuit is that as the thermistor resistance changes the voltage across it changes. This produces a corresponding voltage change across the resistor.

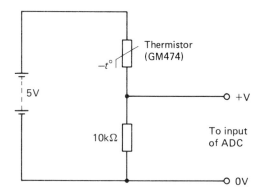

Fig. 12.7 Potential dividing circuit for a thermistor

Because the potential across each component varies with temperature change, the input to the ADC could be taken from either component. Here, the voltage across the resistor is being used to indicate temperature variation.

The basic ADC program can be run and the displayed value (somewhere between 0 and 255) will change as the thermistor temperature changes. This is now a temperature variation indicating system but as yet cannot perform measurements; the system requires calibration. Calibration of such a system was described in Chapter 10.

Figure 12.8 is a typical resistance/temperature curve for a thermistor.* Unfortunately, the characteristics are non-linear and so calibration presents some difficulties.

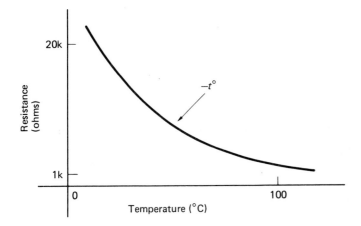

Fig. 12.8 Thermistor characteristics

Due to the decrease in resistance with temperature for the thermistor, the potential across the resistor in the circuit will rise as illustrated in Fig. 12.9.

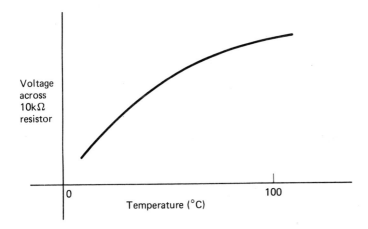

Fig. 12.9

Choice of thermistor and resistor is not critical for the average user. With a poor choice the system will still work but may lack sensitivity.

*Some thermistors have a positive temperature ($+t°$) coefficient i.e. their resistance increases with temperature.

Exercise programs

12.1 Write a traffic light simulation program to include control of the filter lanes. Perhaps you could even include the use of particular keys to simulate traffic flow.

12.2 Re-write the combination lock program so that it incorporates a time delay before the alarm sounds, so that if a genuine mistake is made when entering the combination the entry can be cancelled. This may require the use of a short code for the cancellation procedure.

12.3 Modify program 12.4 so that the number of revolutions completed, and the speed in revs/min are displayed.

12.4 Write software, for the motor control using switches, so that the motor can be 'taught' a set of movements. This sequence can then be repeated time and time again.

Further suggested exercises

Seven segment display

Naturally, many control and monitoring applications involve the display of information, and whether this information is motor speed in revs/min or temperature in °C the microcomputer can show this on the screen. However, in many applications of pure microprocessor control a screen is not available and therefore some other form of display is required. Seven segment displays are one way of providing the information. Such components are used in calculators and digital watches, and shown in Fig. 12.10 is a general type that can be used on printed circuit boards.

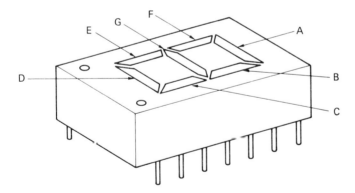

Fig. 12.10 A seven segment display

A small current to a particular segment pin illuminates that particular segment. The current required per segment varies from display to display but is around 10–30mA. To limit the current from a power supply of say 5 volt, resistors are connected to each segment as illustrated in Fig. 12.11. This diagram shows a common anode display but common cathode types are also available.

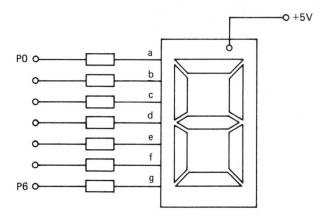

Fig. 12.11 Common anode seven segment display, stage 1

If this type of display is to be used and controlled from the user port then seven lines would be required, leaving only one line free for control of other items. Furthermore, the software must be such that the correct numerical value is displayed, and this is not easy (though it is possible). For example, to display the number one we require illumination of segments B and C and for displaying number two segments A, B, G, E and D must be illuminated. There is no logical sequence therefore, and so the necessary data required to produce the correct value on the display must be stored in the microcomputer, memory and selected when required.

An improvement to this system is to place a decoder prior to the resistors as illustrated in Fig. 12.12. This 7447A decoder chip accepts four bit binary numbers and converts them into the correct form for the seven segment display. This is a much more satisfactory arrangement, as only four user port lines are required, and displaying the correct number simply involves placing the correct binary equivalent on the control lines.

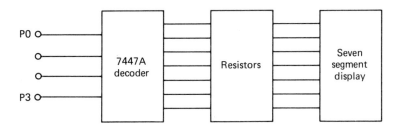

Fig. 12.12 Display and decoder

A further refinement is to include a binary counting chip before the decoder itself (see Fig. 12.13). The output from this chip is the binary equivalent of the number of pulses it has received. Displaying the correct number now is not a software problem and only one line from the user port is required. Obviously using this assembly of components the count is very limited, but a second assembly will provide a count to 99 and in fact the stages can be connected together so that only a single line from the user port is still required.

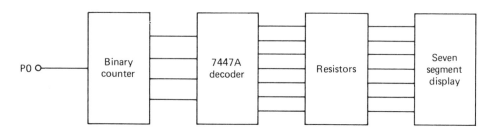

Fig. 12.13 Complete pulse counting arrangement

Strain gauge

There are a whole range of electrical/mechanical components that can be used in conjunction with the microcomputer, but which have not been discussed in this text. One component that has been mentioned though, is the strain gauge and this can be used with a micro to monitor physical changes in a material. Alternatively a simple weighing machine could be produced by placing the gauge on a cantilever arrangement as shown in Fig. 12.14. When a mass is placed on the beam deflection takes place and the strain gauge resistance changes. This change can be monitored with the micro and if the system is calibrated with known masses initially it can be used as a weighing machine.

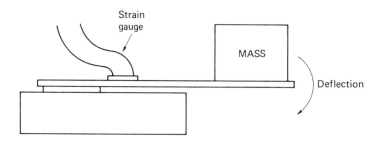

Fig. 12.14 Use of strain gauge

Certain micro-electronics equipment manufacturers produce similar items and the output can be fed directly into an ADC. The devices are usually linear so calibration is simple. The procedure is to run the basic ADC program, place known masses on the beam and record the corresponding VDU display value. A graph can be then drawn and the equation for the line determined (Fig. 12.15). This equation can now be introduced into the basic ADC program and masses displayed directly on the screen. Rearranging the equation below gives (Y–C) / M=X, where X is the mass calculated from Y found at the data register.

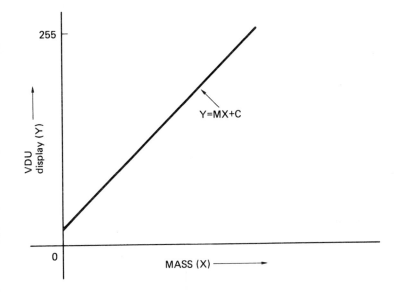

Fig. 12.15 Calibration curve

13

6502 assembly language

This section is aimed at giving the reader a brief and simplified introduction to machine code/assembly language programming. Much of the highly confusing and complicated detail regarding the status register, stack pointer, etc. has been omitted in order to get the reader using the language rather than being bogged down straight away with detail. This omitted information, however, is vital for a real understanding of the language although this can come at a later stage.

Assembly language/machine code is not easy to learn but having said this, just a few hours work and perseverance, and the ice will be cracked and the fear of assembly language will be dispelled. In fact the satisfaction gained, within a short time, by writing a few very simple programmes should be enough to have developed a compelling interest in the language.

What are assembly language and machine code?

Assembly language programs are written in mnemonics. These mnemonics are three letter abbreviations used as assembly language instructions.

Example   LDA #5
This means load a register in the microprocessor known as the accumulator with the number 5. That is, put 5 in the accumulator.

Computers operate which digital signals and these signals can represent binary with numbers. The instruction LDA #5 could be entered by placing the correct binary signal into the computer directly.

Example   10101001 00000101
By entering this string of 1s and 0s we could achieve the same as LDA #5. However, entering all these 1s and 0s would be tedious and prone to errors and so to make life easier hexadecimal numbers are used (see Chapter 2 on number systems). These hexadecimal numbers are the machine codes for the assembly language instructions. The machine code for LDA #5 is A9,05. That is, the machine code for LDA is A9, etc.

Some computers are designed so that assembly language can be used and others can accept machine codes directly. Computers which are programmed in assembly language convert the language directly into machine codes. Printouts of such programs often provide assembly and machine code listings.

Certain microcomputers such as the Commodore 64 and PET have no facility for actually using assembly language initially although the computer is operating with machine codes all the time. It is possible, however, to purchase a chip known as an EPROM (Eraseable Programmable Read Only Memory) which can be plugged directly into the computer's circuit board and can provide access to assembly language use.

Why use assembly language or machine code?

Did you know that the BASIC interpreter is itself a machine code program? Every instruction; IF, THEN, PRINT, GOSUB, etc. is converted into machine codes by the interpreter (or compiler) before it can be executed.

BASIC is classed as a high level language and assembly language as a low level language. However, because machine code is the real language of the computer, writing programs in this language has many advantages over the equivalent BASIC program.

1 Because BASIC instructions have to be converted to machine code, and each BASIC instruction can be a lengthy machine code subroutine, a BASIC program is slow in comparison with an equivalent assembly language program – hundreds of times slower. This is important, especially with control applications and games.

2 Much greater memory economy is achieved than with BASIC for equivalent programs.

3 Learning this low level language provides the student with a much greater insight into the workings of the computer.

Requirements

Before assembly programs can be entered and run on a microcomputer several fundamental facts must be known:

(a) **The type of microprocessor**: the BBC B and PET use the 6502. This means the machine codes will be the same on both machines. The Commodore 64 also uses 6502 machine codes.

(b) **The registers available**: the 6502 microprocessor has three—the accumulator, and the X and Y registers.

(c) **The instruction set** for the microprocessor.

(d) **The safe areas of memory**: with assembly language the programmer decides where in memory things will be stored, but he or she needs to be aware of the memory areas open for use. Not all memory space is available, a great deal of it is taken up on a variety of things such as input/output ports, BASIC routines, etc., etc.

(e) How to get from **BASIC to assembly language and vice versa**. For users of the BBC B the procedure is standard and so presents no real problems, but for users of the PET and Commodore 64 there is no standard technique as this depends on the software purchased for accessing assembly language. However, I will give one typical example of the type of procedure required for the latter two micros. Don't be put off by this, the routines are generally just a line or two of BASIC.

(a) The 6502 microprocessor

Although there are a few slightly different dialects, BASIC is a universal language. Machine code however is not universal. Each make of microprocessor has its own unique machine code. The list of machine codes for the 6502 microprocessor is shown in Table 13.1 (p. 141).

Within this microprocessor (Fig. 13.1) are the accumulator, the X register and the Y register. These are locations where numbers can be stored or manipulated.

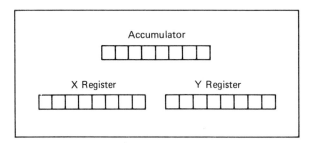

Fig. 13.1 Inside the microprocessor

(b) The accumulator

This is the most important register and it works extremely hard. The accumulator is a general purpose storage area where numbers can be manipulated. Various arithmetic, logical and other strange functions can be carried out by using the accumulator. For example, if two numbers are to be added they are passed to the accumulator, the addition process then takes place and the result is stored in the accumulator.

X and Y registers

These like the accumulator can store numbers, and are very useful registers. However, they cannot perform the broad variety of operations carried out by the accumulator.

The X and Y registers have several useful functions:

1   Either of them can serve as a temporary store of information.

2   An important function is concerned with a technique known as indexed addressing – not discussed in this text.

3   With control programs they provide useful time delays.

(c) The instruction set

The microprocessor instruction set is a list of machine codes and corresponding assembly language mnemonics. The 6502 microprocessor set is shown in Table 13.1 (p. 141).

(d) Safe areas of memory

To determine areas of memory where data can be stored without interfering with other functions one needs to consult a memory map for the microcomputer in question. These can be found in User Guides and text books for the particular micros, and such maps vary considerably in the detail with which they are presented. At this stage you may not require a detailed memory map so I have included an outline for each micro in Appendix 1.

(e) Basic to assembly language and back

Let us examine each microcomputer to see how a simple program can be entered and then run and how to get back to BASIC once this has been done.

BBC B

Entering an assembly language program on the BBC B is similar to entering a BASIC program, except that some extra instructions are required before and after the program.

A simple assembly language program is shown below. Don't worry if at this stage it is meaningless.

```
10 LDA #5
20 STA 5632
30 RTS
```

This three line program cannot simply be entered and run without first indicating to the micro that assembly language is to be used and also where the program is going to be placed in memory. This is achieved with P%=5376. The number 5376 is the location in memory where the program will begin. Alternatively this could be done by P%=&1500, where &1500 is the hexadecimal equivalent of 5376 denary.

Arrows (mode 7) or square brackets are used to enclose the complete program, and the program ready for RUNNING in mode 7 is:

```
 5 P%=5376
 8 ←—— (or [ in other modes)
10 LDA #5
20 STA 5632
30 RTS
40 ——→ (or ] in other modes)
```

On the command RUN the program will be assembled as machine code in memory; starting at 5376. At this stage the program has not been executed and to do this a CALL statement followed by the starting address is typed.

```
CALL 5376
```

The computer will now execute the program and return to BASIC.

This call statement can be included in each program at the end and the complete program ready for execution on the RUN command is:

```
 5 P%=5376
 8 ◄──
10 LDA #5
20 STA 5632
30 RTS
40 ──►
50 CALL 5376
```

If this program is run, nothing apparently happens except that the microcomputer displays a machine code listing for the program (Fig 13.2). Naturally, to understand what the program has achieved one requires to study the language.

(5376 in denary)	Machine codes			Mnemonics
1500				
1500	A9	05		LDA #5
1502	8D	00	16	STA 5632
1505	60			RTS
				(1600 in hex)

Fig. 13.2

Commodore 64 and PET

With these two computers it is not possible to be specific about the technique for entering and running assembly language programs, as is the case with the BBC B. It was mentioned earlier that neither of these two micros initially allow access to programming in assembly language and so the procedure is dictated by the software/EPROM purchased. However, I will outline a system in the hope that some light will be thrown upon the problem.

Supposing an EPROM for providing assembly language programming has been obtained. The instructions with it will explain where it is to be inserted in the circuit board of the microcomputer. Once inserted we are ready to start.

On power up the micro will be in BASIC, or course, and it is necessary now to call up the routine stored in the EPROM. A typical code could be:

SYS 4096*9 | return |

The assembly program can now be entered in just the same way that a BASIC program would be.

It may not be possible to use the RUN command and a typical move now is to assemble the program (as machine coding) into memory by using:

!A   (OR ASSEMBLE) | RETURN |

The program at this stage has simply been positioned in memory and to execute the program it is necessary to indicate to the micro the start address in memory where the program is held; this again is probably dictated by the EPROM. A typical statement used to run the program is:

SYS 826

where 826 is the start address of the program.

To recap, there are only three instructions required:

(1)   SYS 4096*9   calls up the assembly language routine.
                    The program can now be entered.
(2)   !A   assembles program into memory.
(3)   SYS 826   executes program.

Binary/hexadecimal/decimal

The number system to be used throughout this chapter is denary although hexadecimal or binary can generally be used.

Below are a list of prefixes that *may* be used if software/EPROM is used on the Commodore machines.

Example
LDA #10   decimal (no prefix)
LDA #$A   hexadecimal (BBCB is &)
LDA #%101010   binary

Although denary is to be used here, hexadecimal is generally used with assembly language.

6502 assembly language

Learning assembly language is not too difficult if it is done instruction by instruction and by writing a short program using each instruction encountered. One way is to work through the instruction set alphabetically but the most logical way is to start with the most useful instructions first. LDA is the first to be discussed.

LDA (LoaD Accumulator)

This mnemonic means 'load the accumulator with a number'. If we desire to place a number 6 into the accumulator the program line would be:

    30 LDA #6

On running the program the number previously stored in the accumulator is erased and 6 is entered – immediately. This is called **immediate addressing**. The # sign indicates to the assembler immediate addressing.

Diagramatically this looks like this:

accumulator

6 ⟶ [   ]

Learning the first instruction is the most difficult, but hopefully this didn't seem so bad. Although learning one instruction at a time may seem a long and drawn out method, a snowball effect takes place. For example, the next two instructions are very similar to LDA.

LDX/LDY

LDX means load the X register with a number.
LDY means load the Y register with a number.

Example

```
40 LDX #10
50 LDY #5
```

These are again immediate addressing modes so X would be loaded with 10 and Y with 5.

Before proceeding any further let us analyse these statements in terms of machine codes. If you refer to Table 13.1 (p. 141) and look down the left hand column for LDA and then move across the page until you are referring to the column headed IMMEDIATE (addressing) then the number you encounter is A9. This is the machine code for LDA in the immediate addressing mode. Similarly for LDX and LDY in immediate addressing we find A2 and A0 respectively. The column headings shown in Table 13.1 are the different addressing modes available on the 6502 microprocessor. You may wonder, why so many? — well this will become clearer as you progress through this chapter.

Before going onto the next instruction, contrast the following two lines:

Example

```
30 LDA #7
60 LDA 800
```

Line 30 is immediate addressing (#) and loads the accumulator with 7. In line 60 there is no # sign prior to the number (known as the operand) and in fact this is known as **absolute addressing**. This line means load the accumulator with the number presently stored in location 800 in memory, 800 being an absolute address. Naturally, such an addressing mode is essential if data stored in memory is to be manipulated in some way.

Diagramatically this looks as follows:

Memory locations	Value stored	
799	6	
800	12	Accumulator
801	3	12
802	251	

Here the memory location holds the number 12 and so the accumulator is loaded with 12. *Note*—the contents of location 800 remain at 12; the value has not been removed but simply copied.

What do the following instructions mean?

```
 50 LDA 1523
100 LDA #7
110 LDX #6
120 LDY 500
```

STA (STore Accumulator in memory)

This instruction stores the contents of the accumulator into a specified memory location.

Example

```
20 STA 700
```

Here the instruction tells the microprocessor to place that value held in the accumulator into memory location 700. If the accumulator held the number 8 then diagramatically this looks like this:

Memory location	Data
698	
699	
700	8
701	

Accumulator

8 ──────→ [8]

Whatever was in location 700 has been erased and 8 entered. The contents of the accumulator remain unchanged at 8.

The number 8 could be placed into numerous memory locations by:

```
LDA #8
STA 700
STA 701
STA 702
STA 703
```

Memory location	New value
700	8
701	8
702	8
703	8
704	

Accumulator

[8]

This type of instruction is again **absolute addressing**. So what is this type of addressing?

In absolute addressing the operand (the number part of the instruction) is a memory address which can be anywhere in the computer's memory. Numerous instructions utilise this mode of addressing, as can be seen by referring to Table 13.1.

STX/STY

These instructions are used with the X and Y registers in the same way that STA is used with the accumulator.
For example

```
50 STX 850
60 STY 210
```

The value in the X register is placed in location 850 and the value in the Y register is placed in location 210. Again—absolute addressing.

Six instructions have been discussed and not a program in sight. Before moving on to more instructions let us use each of the instructions covered.

PROG. 13.1 Use of LDA and STA

BBC B user note: all the programs require enclosing in arrows or square brackets, etc., as previously described; and all the programs here are designed for MODE 7 use although other modes are perfectly capable of handling them. MODE 7 has been chosen because use of screen addresses in this mode is ideal for certain introductory programs. The micro powers up in MODE 7.

```
        BBC                 COMMODORE 64            PET

   10 LDA #5            10 LDA #5            10 LDA #5
   20 STA 31000         20 STA 50000         20 STA 800
```

This simple program to load and store a number is incomplete. As with BASIC programs the end of the program needs to be specified.

The standard end of program instruction is BRK (Break) so the complete program should look as follows:

```
10 LDA #5
20 STA XXX          ←—— memory location
30 BRK
```

In practice, however, the programs for the BBC B end with the instruction RTS. The programs for Commodore 64 and PET are usually terminated in the same way. Because we are dealing with micros, whose general operating language is BASIC, RTS (ReTurn from Subroutine) is used to return the control to BASIC once the program is executed. The complete program is therefore:

```
        BBC                 COMMODORE 64            PET

   10 LDA #5            10 LDA #5            10 LDA #5
   20 STA 31000         20 STA 50000         20 STA 800
   30 RTS               30 RTS               30 RTS
```

Note the spaces between 10 and LDA and LDA and #5

The program can now be executed although nothing obvious will occur, except on the BBC (and perhaps the Commodore 64 machines) a machine code list appears on the VDU. All we have done is to store a number in a specified memory location, so to see if the program has worked it is necessary to examine the contents of the location. This can be achieved by:

```
      BBC              COMMODORE 64              PET

 PRINT? 31000     PRINT PEEK (50000)     PRINT PEEK (800)
```

5 should be the value displayed, of course.

Try storing other numbers in the same address and then storing numbers in addresses in the vicinity of the addresses given.

Comments

Just as BASIC requires REM statements to provide explanatory notes so assembly programs require explanation. These can be introduced at the end of a program line after a semi-colon.
For example:
10 LDA #20 ;accumulator loaded
20 STA 800   ;value stored at 800
30 RTS   ;end program
            at least one space required

PROG. 13.2 Use of all six instructions

```
      BBC                 COMMODORE 64          PET

10 LDA #10           10 LDA #10           10 LDA #10
20 LDX #3            20 LDX #3            20 LDX #3
30 LDY #4            30 LDY #4            30 LDY #4
40 STX 31000         40 STX 50000         40 STX 801
50 STY 31001         50 STY 50001         50 STY 802
60 STA 31002         60 STA 50002         60 STA 803
70 RTS               70 RTS               70 RTS
```

Run the program and inspect the necessary memory locations using the PEEK instruction (? operator for BBC users).

ASCII codes

The 6502 microprocessor has an 8-line data bus and so can handle numerical data between 0 (binary 00000000) and 255 (binary 11111111). But how does it recognise and deal with non-numeric characters? The answer is: a coding system is used where each letter of the alphabet and characters such as +, #, $ etc., are given a numeric value. The most common coding system is the ASCII code (American Standard Code for Information Interchange). For example the ASCII code for A is 65, B is 66, C is 67, etc. Appendix 2 gives a more comprehensive view of the code. Unfortunately, the Commodore 64 and PET do not use these codes for letters that appear on the screen and in fact for these machines the code for A is 1, B is 2, C is 3, etc.

Using the screen addresses

Because such codings exist we can write assembly programs so that numbers, letters or characters appear exactly where we want them on the screen. A typical monitor can show up to 1000 characters, i.e. it has 25 lines with a possible 40 characters per line. All that is required is the screen addresses. These are the addresses in memory which relate to the screen, and these are given in Fig. 13.3.

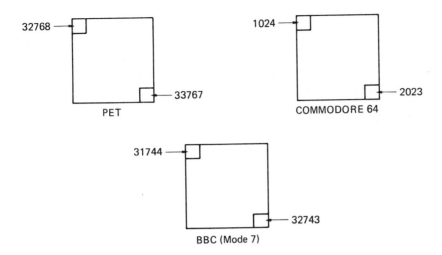

Fig. 13.3 Screen addresses

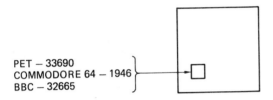

Fig. 13.4

PROG 13.3
This places the letter 'A' on the screen in the position shown in Fig. 13.4.

```
        BBC                 COMMODORE 64            PET

10 LDA #65   ;A        10 LDA #1            10 LDA #1   ,A
20 STA 32666           20 STA 53281         20 STA 33690
30 RTS                 30 LDA #2            30 RTS
                       40 STA 53281
                       50 LDA #1   ;A
                       60 STA 1946
                       70 RTS
```

Before running the program clearing the screen is an advantage so that one can see more clearly what has happened. *Note:* with the BBC B micro the screen address position changes if the information on the screen scrolls. To maintain the screen addresses given in Fig. 13.3 a clear screen is essential (Mode 7 is being used).

The Commodore 64 program seems complicated in comparison with the others and this is because of the screen graphics. If the number 1 is simply loaded to a screen address then the letter A will appear but will be the same colour as the background. The only way to see the letter under these conditions is to place the cursor over the letter when it will be observed. Lines 10 to 40 on the Commodore 64 program layer the screen with colours and now placing the letter 'A' on the screen will produce a white letter on a dark background.

PROG. 13.4 Write your name on the screen using assembly language.

BBC		COMMODORE 64		PET	
10 LDA #83	; S	10 LDA #1		10 LDA #19	; S
20 STA 32666		20 STA 53281		20 STA 33690	
30 LDA #85	; U	30 LDA #2		30 LDA #21	; U
40 STA 32268		35 STA 53281		40 STA 33692	
50 LDA #83	; S	40 LDA #19	; S	50 LDA #19	; S
60 STA 32670		50 STA 1946		60 STA 33694	
70 LDA #65	; A	60 LDA #21	; U	70 LDA #1	; A
80 STA 32672		70 STA 1948		80 STA 33696	
90 LDA #78	; N	80 LDA #19	; S	90 LDA #14	; N
100 STA 32674		90 STA 1950		100 STA 33698	
110 RTS		100 LDA #1	; A	110 RTS	
		110 STA 1952			
		120 LDA #14	; N		
		130 STA 1954			
		140 RTS			

INC (INCrement memory by one)

Example
10 INC 31000 – this would cause the contents of memory location 31000 to be increased by one in the BBC micro.

Such an instruction can be used in a counting loop, say to count up to a certain value. For counting up in twos or more consecutive INC instructions can be used as illustrated in program 13.5.

PROG. 13.5 To use the increment instruction

BBC	COMMODORE 64	PET
10 LDA #0	10 LDA #0	10 LDA #0
20 STA 31000	20 STA 50000	20 STA 800
30 INC 31000	30 INC 50000	30 INC 800
40 INC 31000	40 INC 50000	40 INC 800
50 RTS	50 RTS	50 RTS

The content of the location is set initially at 0. It is then incremented twice so it now holds 2. This can be checked by looking into the location.

By referring to Table 13.1 (p. 141) for the INC instruction one can see that four addressing modes can be used. Which mode was used in program 13.5?

Answer – absolute addressing.

DEC (DECrement memory by one)

This is the opposite of INC. It decrements (decreases) the contents of a specified location by one. Again this instruction can be used in a count loop, say to count down to a certain value.

PROG. 13.6 To use decrement instruction

BBC	COMMODORE 64	PET
10 LDA #3	10 LDA #3	10 LDA #3
20 STA 31500	20 STA 50000	20 STA 810
30 DEC 31500	30 DEC 50000	30 DEC 810
40 DEC 31500	40 DEC 50000	40 DEC 810
50 RTS	50 RTS	50 RTS

Inspection of the contents of the specified location gives a value of 1.

INX/INY

INX and INY allow registers to be incremented by one respectively. Note that no number (operand) is required with these instructions.

PROG. 13.7 Use of INX

BBC	COMMODORE 64	PET
10 LDX #6	10 LDX #6	10 LDX #6
20 INX	20 INX	20 INX
30 INX	30 INX	30 INX
40 INX	40 INX	40 INX
50 STX 31200	50 STX 50005	50 STX 805
60 RTS	60 RTS	60 RTS

Again, we can PEEK into the location to prove this value is actually 9.

One important point to mention is that memory location contents and X and Y register contents can be INCremented but the accumulator cannot be. However, the accumulator value can be incremented by *adding one*. The instruction to achieve this is ADC, but this is not discussed in this text.

DEX/DEY

These are similar, but the reverse of INX and INY. Write a short program to use these two instructions. These instructions will be used a great deal in this text (although INX/INY could be used) in counting loops and provide a very convenient way of constructing time delay routines.

Implied addressing

DEX/DEY/INX/INY are instructions that require no operand—no numbers placed against them. The operand is **implied** rather than specifically stated, and this type of instruction uses what is called **implied addressing** mode.

Reference to Table 13.1, reveals under the IMPLIED column other instructions which operate under the implied addressing mode.

BRK, CLC, CLD, RTS, etc.

So far twelve instructions have been covered but as yet our assembly language vocabulary is not large enough to write *useful* and meaningful programs.

JMP (JUMP to new location)

This is an unconditional jump and it is used like the GOTO statement in BASIC. There are no conditions attached, the program sequence must jump to the specified place.

Consider the program line below:

70 JMP 500

This does not mean jump to line 500—it means jump to memory location 500 and continue the program from there. This is not very convenient to the programmer of the 'home microcomputer'. A more suitable way is to use labels.

Suppose we want the program sequence to jump from line 70 to line 200. Using labels the procedure is:

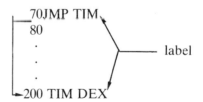

PROG. 13.8 To use the JMP instruction

```
     BBC                    COMMODORE 64            PET

10       LDA #5         10      LDA #5        10      LDA #5
20       STA 31000      20      STA 50000     20      STA 800
30       STA 31001      30      STA 50001     30      STA 801
40       JMP TIM        40      JMP TIM       40      JMP TIM
50       DEC 31001      50      DEC 50001     50      DEC 801
60       DEC 31001      60      DEC 50001     60      DEC 801
70.TIM RTS              70 TIM RTS            70 TIM RTS
```

This program loads two locations (stated in lines 20 and 30) with 5, and then jumps directly to the end of the program without executing the decrement instructions. Inspection of the two locations reveal they both hold 5.

Referring to Table 13.1 again, for the JMP instruction, shows there are two modes of addressing for the JMP instruction. Only the ABSOLUTE mode is being used in this text.

Two-pass assembly

The program above (program 13.8) would no doubt be successful on Commodore 64 and PET machines, as a two-pass assembly routine would probably be a feature of the package bought to access assembly language. The BBC program would fail at line 40 and an error message:

No such variable at line 30

would be displayed.

If a label is defined in a program before it is encountered (i.e. a jump or branch back to a defined label) then no problems arise. If however, as in our program, the jump is to an undefined (at that present time) label TIM then the assembly process halts as the assembler cannot place a value or location on TIM.

So; jumps or branches back to labels are acceptable, whereas jumps or branches forward are unacceptable. To overcome the latter problem on could actually calculate the address or value of the label and define it at the start of the program, but this defeats the object of the labels, providing simplicity. The ideal way, for the BBCB machine, involves two program modifications.
(1)   Using an OPT statement.

OPT 0 assembles errors suppressed, no listing
OPT 1   listing
OPT 2   reported, no listing
OPT 3   listing

These statements are placed after the square brackets (arrows—mode 7).
Using OPT 1 in our BBC program takes us a step nearer to success, and the assembler passes through the entire program as shown on the next page:

```
 5  P% =5400
 7  [ OPT 1
10  LDA £5
20  STA 31000
30  STA 31001
40  JMP TIM
50  DEC 31001
60  DEC 31001
70  .TIM RTS
80  ]
90  CALL 5400
```

```
>RUN
1518              OPT 1
1518 A9 05        LDA £5
151A 8D 18 79     STA 31000
151D 8D 19 79     STA 31001
1520 4C 20 15     JMP TIM
1523 CE 19 79     DEC 31001
1526 CE 19 79     DEC 31001
1529 60           .TIM RTS
```

(2)   Examination of the machine code shows that the value inserted for TIM is wrong. It should be 1529. To achieve the correct value for TIM the program must receive a second pass through the assembler.

This two-pass assembly can be achieved by using a BASIC, FOR . . . NEXT loop routine, as illustrated below:

```
 3  FOR PASS=1 TO 3 STEP 2
 5  P% = 5400
 7  [OPT PASS
10  LDA £5
20  STA 31000
30  STA 31001
40  JMP TIM
50  DEC 31001
60  DEC 31001
70  . TIM RTS
80  ]
85  NEXT PASS
90  CALL 5400
```

```
>RUN
1518              OPT PASS
1518 A9 05        LDA £5
151A 8D 18 79     STA 31000
151D 8D 19 79     STA 31001
1520 4C 20 15     JMP TIM
1523 CE 19 79     DEC 31001
1526 CE 19 79     DEC 31001
1529 60           . TIM RTS
1518              OPT PASS
1518 A9 05        LDA £5
151A 8D 18 79     STA 31000
151D 8D 19 79     STA 31001
1520 4C 29 15     JMP TIM
1523 CE 19 79     DEC 31001
1526 CE 19 79     DEC 31001
1529 60           .TIM RTS
```

In this routine the variable PASS takes the value of 1 for the first pass and 3 for the second pass and so error messages are suppressed in the first pass only.

JSR/RTS

JSR—jump to new location saving return address (Jump to SubRoutine)

RTS—ReTurn from Subroutine.

JSR and RTS are used in just the same way as GOSUB and RETURN in BASIC. Just as GOSUB requires a line number for the subroutine (GOSUB 500), JSR requires either the location in memory of the subroutine (JSR 795) or a label (JSR DEL). As in BASIC, on the return from the subroutine the program carries on from where it left off.

Subroutines are used commonly in control applications for time delays which may be used several times in a control sequence, for example stepper motor control or traffic light simulators.

PROG. 13.9
This uses a subroutine to place letters A, B, and C on the screen.

	BBC			COMMODORE 64			PET	
10	LDX	#64	10	LDX	#1	10	LDX	#0
20	JSR	TIM	20	STA	53281	20	JSR	TIM
30	STX	32000	30	LDA	#2	30	STX	33000 ;A
40	JSR	TIM	40	STA	53281	40	JSR	TIM
50	STX	32000	50	LDX	#0	50	STX	33002 ;B
60	JSR	TIM	60	JSR	TIM	60	JSR	TIM
70	STX	32000	70	STX	1948	70	STX	33004 ;C
80	RTS		80	JSR	TIM	80	RTS	
200	.TIM INX		90	STX	1950	200	TIM INX	
210	RTS		100	JSR	TIM	210	RTS	
			110	STX	1952			
			120	RTS				
			200	TIM	INX			
			210	RTS				

What type of addressing is used for RTS and JSR?

Branch instructions

There are eight branch instructions in the 6502 instruction set. These are conditional branches, that is, a branch occurs provided certain conditions are satisfied. IF X =   THEN, is the equivalent type of instruction in BASIC.

BEQ, BMI, BNE, BPL are the four most useful to the newcomer to assembly language and these are the only ones to be discussed.

BEQ   Branch if equal to zero
BNE   Branch if not equal to zero
BPL   Branch if plus
BMI   Branch if minus

To appreciate these instructions fully requires knowledge of the status register. Initially, however, it is best to use the instructions by simply remembering their BASIC equivalent.

Branch instructions	BASIC equivalent
BEQ	IF A = 0 THEN . . . .
BNE	IF A <> 0 THEN . . . .
BPL	IF A < 128 THEN . . . .
BMI	IF A > = 128 THEN . . . .

The last two conditions seem to have little to do with plus or minus and require a knowledge of negative binary numbers to be fully understood. However, my initial aim is practical rather than theoretical and lack of this information in no way detracts from the application of these conditions.

So branch instructions exist, but branch if what? The implication is, branch if the last result was zero (not zero, etc.)

Branches can be forwards or backwards in a program as shown in Fig. 13.5.

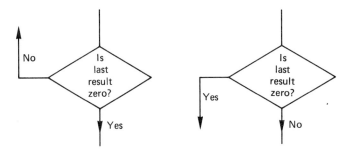

Fig. 13.5 Branches

Relative addressing

This form of addressing is applied only to conditional branch instructions. The branch is **relative** to the present position in the program and to understand this fully consider the program section below.

Memory location	Machine code	Assembly language
0032	85,60	STA 60
0034	88	DEY
0035	D0,FD	BNE −3
0037	C6,60	DEC 61

The instruction BNE −3 would cause a branch three bytes backwards if the condition was satisfied. If the condition was not satisfied the instruction would be passed over and the DEC instruction would be executed.

So it is important for us to remember that an instruction such as

50 BNE 09

means jump forward by nine memory locations (or bytes) if the condition is satisfied and NOT nine lines in the program.

Fortunately labels can be used on many microcomputers which saves much inconvenience. The same instruction might read:

```
┌─50      BNE L1
│ 60
│ 70
│ 80
│ 90
└►100     L1 LDA #10
```

PROG. 13.10 To decrement the contents of a specified location to zero.

	BBC		COMMODORE 64			PET	
10	LDA #20	10	LDA #20	10		LDA #20	
20	STA 31000	20	STA 50000	20		STA 800	
30 .L1	DEC 31000	30 L1	DEC 50000	30	L1	DEC 800	
40	BNE L1	40	BNE L1	40		BNE L1	
50	RTS	50	RTS	50		RTS	

To see if the program has worked PEEK (or ?) into memory.

Time delays

Branch instructions can be used in conjunction with decrement (or increment) instructions to build useful time delay routines. The actual time delay can be calculated by the programmer because the precise time taken for the microprocessor to perform each instruction is known. A list is shown in Table 13.2 (p. 144). For example the time to decrement the X register (DEX) takes two clock cycles.

The microprocessor operates at a speed of 1MHz (1000000 cycles/second). It takes 1/1000000th (0.000001) of a second for each action it performs. These actions are performed each time it receives a pulse generated by a so called clock chip. The speed of the clock chip determines the speed of the microprocessor.

Very short delay—single loop

The X (or Y) register is loaded with 255 then decremented to zero. BNE continues to decrement loop until zero is reached as indicated by the arrow.

```
   500     LDX #255     ;load up 255
┌▶510 L1 DEX
└─520       BNE L1       ;is X = 0?
```

The precise time for this delay can be calculated fairly easily but it is approximately one thousandth of a second—1ms (1 millisecond). This delay is far too short for many applications, such as traffic light control, so a double loop is required. This delay is very similar to the BASIC counting loops such as

FOR K = 1 TO 20:NEXT K.

$\simeq \frac{1}{2}$ **second time delay—double loop**

A double loop, that is, a loop inside a loop produces a delay in the region of $\frac{1}{2}$ second.

```
     500     LDY #255
  ┌▶510 L2 LDX #255
┌─┤ 520 L1 DEX
│ └─ 530       BNE L1       ;is X zero
│    540       DEY
└──── 550       BNE L2       ;is Y zero
```

Both the X and Y registers are loaded with 255. The X register is then decremented to zero without touching the Y register.
The Y register is now decremented by 1 to 254.
The X register is again loaded with 255 and decremented to zero.
The Y register is now decremented by 1 to 253.
And so the process goes on.

The X register is decremented from 255 to 0 two hundred and fifty five times. This means the time delay is 255 times greater than the previous delay.
  Note that the microprocessor decrements the X register 255×255 times, approximately 65000 in about half a second. To count to 65000 in BASIC would take about one and half minutes on the Commodore micros. This can be proved by running the following BASIC program and timing your computer:

```
10 PRINT "STARTED"
20 FOR I = / TO 65000
30 NEXT I
40 PRINT "FINISHED"
```

Although we appear to have made the microprocessor work extremely hard the delay is still not enough for many applications and a triple loop is required.

Triple loop time delay
The X and Y registers can be used to form two loops but a third loop will have to be formed either by subtracting one from the accumulator or decrementing the contents of a memory location. Using the latter the delay looks like this:

```
500       LDA #8
510       STA 800      ;load up 8
520 L3 LDY #255
530 L2 LDX #255
540 L1 DEX
550       BNE L1       ;is X zero
560       DEY
570       BNE L2       ;is Y zero
580       DEC 800
590       BNE L3       ;is 800 zero
```

The previous double loop is repeated 8 time here. Generally speaking triple loops, such as this, provide satisfactory delays for most applications. Decrement instructions have been used but increments work just as satisfactorily.

One suggestion – it is a great advantage, and time saver, to load the delay onto floppy disc, and re-load it prior to writing a new program that will require a delay.

PROG. 13.11
This places the letter 'A' on the screen about half a second before the letter 'B' appears,

BBC			COMMODORE 64			PET		
10	LDA #65	;A	10	LDA #1		10	LDA #1	;A
20	STA 32000		20	STA 53281		20	STA 33000	
30	LDX #255		30	LDX #2		30	LDX #255	
40.L2	LDY #255		40	STA 53281		40 L2	LDY #255	
50.L1	DEY		50	LDA #1	;A	50 L1	DEY	
60	BNE L1		60	STA 1946		60	BNE L1	
70	DEX		70	LDX #255		70	DEX	
80	BNE L2		80 L2	LDY #255		80	BNE L2	
90	LDA #66	;B	90 L1	DEY		90	LDA #2	;B
100	STA 32002		100	BNE L1		100	STA 33002	
110	RTS		110	DEX		110	RTS	
			120	BNE L2				
			130	LDA #2	;B			
			140	STA 1948				
			150	RTS				

PROG. 13.12
This prints your name on the screen with letters appearing at about three second intervals (refer back to program 13.4).

	BBC		COMMODORE 64		PET	
10	LDA #83	10	LDA #1	10	LDA #19 ;S	
20	STA 32666	20	STA 53281	20	STA 32990	
30	JSR DEL	30	LDA #2	30	JSR DEL	
40	LDA #85	40	STA 53281	40	LDA #21 ;U	
50	STA 32668	50	LDA #19	50	STA 32992	
60	JSR DEL	60	STA 1946	60	JSR DEL	
70	LDA #83	70	JSR DEL	70	LDA #19 ;S	
80	STA 32670	80	LDA #21	80	STA 32994	
90	JSR DEL	90	STA 1948	90	JSR DEL	
100	LDA #65	100	JSR DEL	100	LDA #1 ;A	
110	STA 32672	110	LDA #19	110	STA 32996	
120	JSR DEL	120	STA 1950	120	JSR DEL	
130	LDA #78	130	JSR DEL	130	LDA #14 ;N	
140	STA 32674	140	LDA #1	140	STA 32998	
150	JSR DEL	150	STA 1952	150	RTS	
160	RTS	160	JSR DEL	500 DEL	LDA #7	
500 .DEL	LDA #7	170	LDA #14	510	STA 800	
510	STA 800	180	STA 1954	520 L3	LDY #255	
520 .L3	LDY #255	190	JSR DEL	530 L2	LDX #255	
530 .L2	LDX #255	200	RTS	540 L1	DEX	
540 .L1	DEX	500 DEL	LDA #7	550	BNE L1	
550	BNE L1	510	STA 800	560	DEY	
560	DEY	520 L3	LDY #255	570	BNE L2	
570	BNE L2	530 L2	LDX #255	580	DEC 800	
580	DEC 800	540 L1	DEX	590	BNE L3	
590	BNE L3	550	BNE L1	600	RTS	
600	RTS	560	DEY			
		570	BNE L2			
		580	DEC 800			
		590	BNE L3			
		600	RTS			

If you can now follow a program of this length you must be feeling an air of satisfaction.

CMP (Compare memory with accumulator)

The procedure with this instruction is to compare a value with the value stored in the accumulator. This is done by subtracting the comparison value from the accumulator (accumulator − memory value). The result of the subtraction can obviously be zero (if both are equal), positive or negative, and it is these outcomes which are used and not the numerical outcome. Note that the contents of the accumulator and memory remain unchanged.

Branch instructions can be used directly after compare instructions. The branch instructions only allow a test of zero, not zero, plus or minus but not specific values, but when used with CMP any acceptable value can be tested for.

Both immediate and absolute addressing can be used with CMP, for example:

(a) Immediate addressing

```
80     LDA 151    ;load accumulator for location 151
90     CMP #83    ;compare accumulator with 83
100    BEQ L1     ;branch on result zero
.
.
.
150 L1 DEX
```

(b) Absolute addressing

```
180 L2 LDY #50
190
200
210    LDA 151    ;load accumulator from location 151
220    CMP 800    ;compare accum. value with value in 800 location
230    BEQ L2     ;branch on result zero
```

CPX/CPY

These instructions are used in exactly the same way as CMP except that here the comparison is with either the X or Y registers; not the accumulator.

PROG. 13.13 To increment the value of a location to 55

```
        BBC

10        LDA #0       ;load up with 0
20        STA 31000    ;store 0 in location 31000
30 .L1 INC 31000       ;increment value by 1
40        LDA 31000    ;load accumulator ready for comparison
50        CMP #55      ;compare accumulator with 55
60        BEQ L2       ;is result zero
70        JMP L1       ;jump back to increment again, IF NOT ZERO
80 .L2 RTS             ;end of program
```

```
        PET                          COMMODORE 64

10        LDA #0             10        LDA #0
20        STA 800            20        STA 50000
30 L1 INC 800               30 L1 INC 50000
40        LDA 800            40        LDA 50000
50        CMP #55            50        CMP #55
60        BEQ L2             60        BEQ L2
70        JMP L1             70        JMP L1
80 L2 RTS                   80 L2 RTS
```

TAX/TXA/TAY/TYA (transfers between registers)

These instructions provide transfers of data between registers.

TAX   Transfer accumulator value to X register
TXA   Transfer X register value to accumulator
TAY   Transfer accumulator value to Y register
TYA   Transfer Y register value to accumulator

These transfers are generally used to shuffle around data that is in the way, but too important to simply erase.

PROG 13.14 To use TAX and TAY instructions

BBC

```
10      LDA #50         ;Load up with 50
20      STA 31000       ;store 50 in location 800
30      TAX             ;transfer 50 to X
40      TAY             ;transfer 50 to Y
50      STX 31001       ;store X
60      STY 31002       ;store Y
70      RTS
```

```
COMMODORE 64                        PET

10      LDA #50             10      LDA #50
20      STA 50000           20      STA 800
30      TAX                 30      TAX
40      TAY                 40      TAY
50      STX 50001           50      STX 801
60      STY 50002           60      STY 802
70      RTS                 70      RTS
```

Inspection of the relevant locations shows the program has worked.

Try a few programs loading up numbers into the X and Y registers and accumulator and use the transfer instructions.

Logical instructions

When BASIC is used as the control language, bits can be set (or cleared) to switch on (or switch off) some devices without affecting the state of the other control lines by using logical instructions. Logical instructions are also used in assembly language control situations to achieve the same result.

The following rules apply:

AND—used to clear (set at 0) selected bits by using zeros in the mask. Devices can be switched off.

ORA—used to set (set at 1) selected bits by using ones in the mask. Devices can be switched on.

Mask

The term **mask** is used to describe the pattern of bits required to change particular bits in another byte.

AND instruction

This performs a logical AND between a number and the accumulator. Reference to Table 13.1 shows that numerous addressing modes are available for this instruction. Only the **immediate** and **absolute** modes will be used here

Example

Let us suppose that the LED control circuit was being used and that at the present time all eight LEDs were on. If we require to switch off the LEDs on say P7(128) and P0(1) then this could be achieved as follows:

```
100    LDA 65120    here the DR for the BBC is
110    AND #126     being used but the PET or
120    STA 65120    Commodore 64 apply equally well
```

Line 100 loads the accumulator with the value held in the data register, and because all the LEDs are on, 255 will be loaded. Line 110 is the MASK required to clear bits P0 and P7. The result of this action is placed in the accumulator.

Accumulator originally 11111111
And #126 (MASK)   01111110

Result in accumulator   01111110

Line 120 re-stores the accumulator value into the data register and the LEDs on P0 and P7 are switched off.

Example

Suppose that the accumulator holds 15 and memory location 8000 holds 232. If the instruction shown below is used, determine the new value in the accumulator.

80 AND 8000

Solution

Accumulator originally   00001111 = 15
AND 8000   11101000 = 232

New value in accumulator 00001000 = 8

PROG. 13.15 Use of AND instruction

	BBC		COMMODORE 64		PET
10	LDA #254	10	LDA #254	10	LDA #254
20	AND #253	20	AND #253	20	AND #253
30	STA 31000	30	STA 50000	30	STA 805
40	RTS	40	RTS	40	RTS

Line 10 loads accumulator with   11111110 (254)
Line 20 ANDs accumulator with   11111101 (253)

Line 30 stores the result in memory   11111100 (252)
Inspection of the relevant memory location should give the value 252.

ORA instruction

This performs a logical OR between a number and the accumulator, the result of which is stored in the accumulator.

With this instruction as with the logical AND several modes of addressing are available, but only the immediate and absolute are used in this text.

Example

If the LED control circuit was being used and only four LEDs P0 to P3 were on, and we wish to switch on the LED on P7(128), then this could be done as follows:

```
140    LDA 56577   Commodore 64 data register is being
150    ORA #128    used here
160    STA 56577
```

Line 140 loads the accumulator with the value held in the data register—this being 15.
Line 150 is the MASK used to set the P7(128) bit, to switch on the LED. The result of this action is placed in the accumulator.
Accumulator originally   00001111
ORA #128 (Mask)   10000000

Result in accumulator   10001111
Line 160 re-stores the contents of the accumulator into the data register and now five LEDs are on.

Example

If the accumulator holds 7 initially and memory location 5300 holds 128, determine the result in the accumulator if the following instruction is used:

Table 13.1 Summary of instruction codes and addressing modes

	Accumulator	Immediate	Zero Page	Zero Page, X	Zero Page Y	Absolute	Absolute, X	Absolute, Y	Implied	Relative	(Indirect, X)	(Indirect), Y	Absolute Indirect
ADC		69	65	75		6D	7D	79			61	71	
AND		29	25	35		2D	3D	39			21	31	
ASL	0A		06	16		0E	1E						
BCC										90			
BCS										B0			
BEQ										F0			
BIT		2C											
BMI										30			
BNE										D0			
BPL										10			
BRK									00				
BVC										50			
BVS										70			
CLC									18				
CLD									D8				
CLI									58				
CLV									B8				
CMP		C9	C5	D5		CD	DD	D9			C1	D1	
CPX		E0	E4			EC							
CPY		C0	C4			CC							
DEC			C6	D6		CE	DE						
DEX									CA				
DEY									88				
EOR		49	45	55		4D	5D	59			41	51	
INC			E6	F6		EE	FE						
INX									E8				
INY									C8				
JMP						4C							6C
JSR						20							
LDA		A9	A5	B5		AD	BD	B9			A1	B1	
LDX		A2	A6		B6	AE		BE					
LDY		A0	A4	B4		AC	BC						
LSR	4A		46	56		4E	5E						
NOP									EA				
ORA		09	05	15		0D	1D	19			01	11	
PHA									48				
PHP									08				
PLA									68				
PLP									28				
ROL	2A		26	36		2E	3E						
ROR	6A		66	76		6E	7E						
RII									40				
RTS									60				
SBC		E9	E5	F5		ED	FD	F9			E1	F1	
SEC									38				
SED									F8				
SEI									78				
STA			85	95		8D	9D	99			81	91	
STX			86		96	8E							
STY			84	94		8C							
TAX									AA				
TAY									A8				
TYA									98				
TSX									BA				
TXA									8A				
TXS									9A				

70 ORA 5300

Solution

| Accumulator originally | 00000111 | = | 7 |
| ORA 5300 | 10000000 | = | 128 |

New value in accumulator 10000111 = 135

PROG. 13.16

	BBC			COMMODORE 64			PET
10	LDA #15	10	LDA #15		10	LDA #15
20	ORA #240	20	ORA #240		20	ORA #240
30	STA 800	30	STA 50000		30	STA 31000
40	RTS	40	RTS		40	RTS

Line 10 loads accumulator with   00001111   (15)
Line 20 ORs accumulator with   11110000   (240)

Line 30 stores result in memory   11111111   (255)

The usefulness of logical instructions may not be clear at present but if you can follow the mathematics of them, using them for control purposes should provide no real difficulty. The following chapter shows how assembly language can be used for control and I have included as many logical instructions as possible with the introductory control applications discussed. If you examine closely the previous examples on logical instructions, it will become clear that bits can be set by using ORs and bits can be cleared by using ANDs.

Zero page addressing

This is an important aspect of assembly language programming, but because it is not always available for use on microcomputers, I will simply explain what is meant by zero page, rather than going into the use of zero page addressing.

Consider a computer with a 1k memory. In decimal this is 0 to just over 1000. If hexadecimal numbers are used then the memory can be split into sections called **pages.** For example:

	Memory location in hex	Memory location in dec
Page 0	0000–00FF	0–255
Page 1	0100–01FF	256–511
Page 2	0200–02FF	512–
Page 3	0300–03FF	–etc.
	Page number	

The numbers 0,1,2,3 in the hexadecimal numbers indicate the page numbers.

Page zero is 0000 to 00FF in hexadecimal, i.e. 0 to 255 in decimal. Its use is just like using absolute addressing but I would prefer not to go into further detail at this time.

Having covered about 50% of the assembly language instructions we are now in a position to use this language to control devices such as stepper motors, traffic light simulators, etc. There are, however, many important instructions unrevealed and I suggest you now tackle a comprehensive text on assembly language.

Table 13.2(a)   Microprocessor instruction set – alphabetic sequence

ADC	Add Memory to Accumulator with Carry	JSR	Jump to New Location Saving Return Address
AND	'AND' Memory with Accumulator	LDA	Load Accumulator with Memory
		LDX	Load index X with Memory
BCC	Branch on Carry Clear	LDY	Load index Y with Memory
BCS	Branch on Carry Set	LSR	Shift Right One Bit (Memory of Accumulator)
BEQ	Branch on Result Zero		
BIT	Test Bits in Memory with Accumulator	NOP	No Operation
BMI	Branch on Result Minus		
BNE	Branch on Result not Zero	ORA	'OR' Memory with Accumulator
BPL	Branch on Result Plus	PHA	Push Accumulator on Stack
BRK	Force Break	PHP	Push Processor Status on Stack
BVC	Branch on Overflow Clear	PLA	Pull Accumulator from Stock
BVS	Branch on Overflow Set	PLP	Pull Processor Status from Stack
CLC	Clear Carry Flag	ROL	Rotate One Bit Left (Memory or Accumulator)
CLD	Clear Decimal Mode	ROR	Rotate One Bit Right (Memory or Accumulator)
CLI	Clear Interrupt Disable Bit	RTI	Return from interrupt
CLV	Clear Overflow Flag	RTS	Return from Subroutine
CMP	Compare Memory and Accumulator		
CPX	Compare Memory and Index X	SBC	Subtract Memory from Accumulator with Borrow
CPY	Compare Memory and Index Y	SEC	Set Carry Flag
		SED	Set Decimal Mode
DEC	Decrement Memory by One	SEI	Set Interrupt Disable Status
DEX	Decrement Index X by One	STA	Store Accumulator in Memory
DEY	Decrement Index Y by One	STX	Store Index X in Memory
		STY	Store Index Y in Memory
EOR	'Exclusive-Or' Memory with Accumulator		
		TAX	Transfer Accumulator to Index X
INC	Increment Memory by One	TAY	Transfer Accumulator to Index Y
INX	Increment Index X by One	TSX	Transfer Stack Pointer to Index X
INY	Increment Index Y by One	TXA	Transfer Index X to Accumulator
		TXS	Transfer Index X to Stack Pointer
JMP	Jump to New Location	TYA	Transfer Index Y to Accumulator

Table 13.2(b)   Instruction addressing modes and related execution times (in clock cycles)

	Accumulator	Immediate	Zero Page	Zero Page, X	Zero Page, Y	Absolute	Absolute, X	Absolute, Y	Implied	Relative	(Indirect, X)	(Indirect), Y	Absolute Indirect
ADC	.	2	3	4	.	4	4*	4*	.	.	6	5*	.
AND	.	2	3	4	.	4	4*	4*	.	.	6	5*	.
ASL	2	.	5	6	.	6	7	.	.	.	.	.	.
BCC	.	.	.	.	.	.	.	.	.	2**	.	.	.
BCS	.	.	.	.	.	.	.	.	.	2**	.	.	.
BEQ	.	.	.	.	.	.	.	.	.	2**	.	.	.
BIT	3*	.	3	.	.	4	.	.	.	.	.	.	.
BMI	.	.	.	.	.	.	.	.	.	2**	.	.	.
BNE	.	.	.	.	.	.	.	.	.	2**	.	.	.
BPL	.	.	.	.	.	.	.	.	.	2**	.	.	.
BRK	.	.	.	.	.	.	.	.	.	.	.	.	.
BVC	.	.	.	.	.	.	.	.	.	2**	.	.	.
BVS	.	.	.	.	.	.	.	.	.	2**	.	.	.
CLC	.	.	.	.	.	.	.	.	2	.	.	.	.
CLD	.	.	.	.	.	.	.	.	2	.	.	.	.
CLI	.	.	.	.	.	.	.	.	2	.	.	.	.
CLV	.	.	.	.	.	.	.	.	2	.	.	.	.
CMP	.	2	3	4	.	4	4*	4*	.	.	6	5*	.
CPX	.	2	3	.	.	4	.	.	.	.	.	.	.
CPY	.	2	3	.	.	4	.	.	.	.	.	.	.
DEC	.	.	5	6	.	6	7	.	.	.	.	.	.
DEX	.	.	.	.	.	.	.	.	2	.	.	.	.
DEY	.	.	.	.	.	.	.	.	2	.	.	.	.
EOR	.	2	3	4	.	4	4*	4*	.	.	6	5*	.
INC	.	.	5	6	.	6	7	.	.	.	.	.	.
INX	.	.	.	.	.	.	.	.	2	.	.	.	.
INY	.	.	.	.	.	.	.	.	2	.	.	.	.
JMP	.	.	.	.	.	3	.	.	.	.	.	.	5
JSR	.	.	.	.	.	6	.	.	.	.	.	.	.
LDA	.	2	3	4	.	4	4	4*	.	.	6	5*	.
LDX	.	2	3	.	4	4	.	4*	.	.	.	.	.
LDY	.	2	3	4	.	4	4*	.	.	.	.	.	.
LSR	2	.	5	6	.	6	7	.	.	.	.	.	.
NOP	.	.	.	.	.	.	.	.	2	.	.	.	.
ORA	.	2	3	4	.	4	4*	4*	.	.	6	5*	.
PHA	.	.	.	.	.	.	.	.	3	.	.	.	.
PHP	.	.	.	.	.	.	.	.	3	.	.	.	.
PLA	.	.	.	.	.	.	.	.	4	.	.	.	.
PLP	.	.	.	.	.	.	.	.	4	.	.	.	.
ROL	2	.	5	6	.	6	7	.	.	.	.	.	.
ROR	2	.	5	6	.	6	7	.	.	.	.	.	.
RTI	.	.	.	.	.	.	.	.	6	.	.	.	.
RTS	.	.	.	.	.	.	.	.	6	.	.	.	.
SBC	.	2	3	4	.	4	4*	4*	.	.	6	5*	.
SEC	.	.	.	.	.	.	.	.	2	.	.	.	.
SED	.	.	.	.	.	.	.	.	2	.	.	.	.
SEI	.	.	.	.	.	.	.	.	2	.	.	.	.
STA	.	.	3	4	.	4	5	5	.	.	6	6	.
STX	.	.	3	.	4	4	.	.	.	.	.	.	.
STY	.	.	3	4	.	4	.	.	.	.	.	.	.
TAX	.	.	.	.	.	.	.	.	2	.	.	.	.
TAY	.	.	.	.	.	.	.	.	2	.	.	.	.
TSX	.	.	.	.	.	.	.	.	2	.	.	.	.
TXA	.	.	.	.	.	.	.	.	2	.	.	.	.
TXS	.	.	.	.	.	.	.	.	2	.	.	.	.
TYA	.	.	.	.	.	.	.	.	2	.	.	.	.

* Add one cycle if indexing across page boundary
**Add one cycle if branch is taken. Add one additional if branching operation crosses page boundary

14

Control using assembly languages

LED control circuit

The first device to be controlled with software written in BASIC was the LED control circuit, and here again it will prove to be the most useful item. As was the case with BASIC control the first program using assembly language is simply to switch on the desired LEDs. This was achieved previously with:

```
        BBC                 COMMODORE 64              PET

10  ?65122=255       10  POKE 56579,255      10  POKE 59459,255
20  ?65120=X         20  POKE 56577,X        20  POKE 59471,X
30  END              30  END                 30  END
```

Where X is any number between 0 and 255 and dictates which LEDs will be on. To produce the equivalent program in assembly language we must follow the same line of action. That is, the DDR must be set to begin with, and this can be achieved by:

```
        BBC                 COMMODORE 64              PET

10  LDA #255          10  LDA #255            10  LDA #255
20  STA 65122         20  STA 56579           20  STA 59459
```

In other words: POKE = LDA + STA,

The complete program follows.

PROG. 14.1 LED Control

```
        BBC                 COMMODORE 64              PET

10  LDA #255          10  LDA #255            10  LDA #255
20  STA 65122         20  STA 56579           20  STA 59459
30  LDA #X            30  LDA #X              30  LDA #X
40  STA 65120         40  STA 56577           40  STA 59471
50  RTS               50  RTS                 50  RTS
```

Lines 10 and 20 set the DDR so that all the control lines are outputs.

Lines 30 and 40 set the bits in the DR, therefore switching on particular LEDs, X being a number in the range 0 to 255.

The second program, when BASIC was used, was to switch all the LEDs off after a period of time and the equivalent in assembly language is given in program 14.2.

PROG. 14.2 Switching LEDs off after a short time delay

	BBC			COMMODORE 64			PET	
10		LDA #255	10		LDA #255	10		LDA #255
20		STA 65122	20		STA 56579	20		STA 59459
30		LDA #X	30		LDA #X	30		LDA #X
40		STA 65120	40		STA 56577	40		STA 59471
50		LDY #255	50		LDY #255	50		LDY #255
60	.L2	LDX #255	60	L2	LDX #255	60	L2	LDX #255
70	.L1	DEX	70	L1	DEX	70	L1	DEX
80		BNE L1	80		BNE L1	80		BNE L1
90		DEY	90		DEY	90		DEY
100		BNE L2	100		BNE L2	100		BNE L2
110		LDA #0	110		LDA #0	110		LDA #0
120		STA 65120	120		STA 56577	120		STA 59471
130		RTS	130		RTS	130		RTS

Lines 10 and 20 set all control lines as outputs.

Lines 30 and 40 dictate which LEDs are illuminated.

Lines 50 to 100 provide double loop time delay.

Lines 110 and 120 switch all the LEDs off.

Following the procedure used for control of the LEDs with BASIC, the next step is to produce a program to switch the LEDs on and off continuously. As with BASIC, a subroutine houses the time delay. One problem arises here for the users of Commodore 64 and PET micros, and that is that the software produces a continuous program loop, and escape from this is apparently impossible, and the only answer is to switch the micro off and start again. In fact by using memory locations 197 (Commodore 64) and 151 (PET) particular keys can be used to escape from the loop, as can be seen by an examination of program 14.3. The BBC B fortunately does not have this shortcoming. These locations (151 and 197) were discussed in Chapter 5 (program 5.3). The keys chosen for escape are S (PET) and T (Commodore 64).

PROG. 14.3 Flashing LEDs

	BBC			COMMODORE 64			PET	
10		LDA #255	10		LDA #255	10		LDA #255
20		STA 65122	20		STA 56579	20		STA 59459
30	.BEG	LDA #15	30	BEG	LDA #15	30	BEG	LDA #15
40		STA 65120	40		STA 56577	40		STA 59471

```
50    ————          50    LDA 197         50    LDA 151
60    ————          60    CMP #22         60    CMP #83
70    ————          70    BEQ FIN         70    BEQ FIN
80    JSR DEL        80    JSR DEL         80    JSR DEL
90    LDA #0         90    LDA #0          90    LDA #0
100   STA 65120      100   STA 56577       100   STA 59471
110   JSR DEL        110   JSR DEL         110   JSR DEL
120   JMP BEG        120   JMP BEG         120   JMP BEG
130   ————          130 FIN RTS           130 FIN RTS
500 .DEL LDX #255    500 DEL LDX #255      500 DEL LDX #255
510 .L2 LDY #255     510  L2 LDY #255      510  L2 LDY #255
520 .L1 DEY          520  L1 DEY           520  L1 DEY
530    BNE L1        530    BNE L1         530    BNE L1
540    DEX           540    DEX            540    DEX
550    BNE L2        550    BNE L2         550    BNE L2
560    RTS           560    RTS            560    RTS
```

Lines 10 and 20 set all control lines as outputs.
Lines 30 and 40 produce output to the LEDs.
Lines 50 to 70 provide an escape (to line 130) from the loop (for Commodore 64 and PET only).
Lines 80 and 110 bring in the time delay starting at line 500.
Lines 90 and 100 switch off all the LEDs.

Remember— Commodore 64 users press *T* to escape.

PET users press S to escape.

Logical instructions

A very important feature of control software, as was previously mentioned, is the ability to set/clear particular bits and test for bits set or clear. Just as in BASIC, assembly language uses logical instructions to do this, and the same fundamental principles apply. To set a bit we use an OR (written as ORA) instruction and to clear a bit we use an AND instruction, etc.

Program 14.4 initially illuminates four LEDs (P0/P3), but an ORA instruction is used to switch the LED on P7.

PROG. 14.4 Use of ORA instruction

```
   BBC              COMMODORE 64          PET

10 LDA #255        10 LDA #255         10 LDA #255
20 STA 65122       20 STA 56579        20 STA 59459
30 LDA #15         30 LDA #15          30 LDA #15
40 STA 65120       40 STA 56577        40 STA 59471
50 ORA #128        50 ORA #128         50 ORA #128
60 STA 65120       60 STA 56577        60 STA 59471
70 RTS             70 RTS              70 RTS
```

By using the same general program, but replacing the ORA with an AND instruction we can switch off a particular LED without affecting the others.

PROG. 14.5 Use of AND instruction

```
        BBC                    COMMODORE 64              PET

10 LDA #255              10 LDA #255              10 LDA #255
20 STA 65122            20 STA 56579            20 STA 59459
30 LDA #15              30 LDA #15              30 LDA #15
40 STA 65120            40 STA 56577            40 STA 59471
50 AND #251            50 AND #1251           50 AND #251
60 STA 65120            60 STA 56577            60 STA 59471
70 RTS                  70 RTS                  70 RTS
```

With this program, initially the LEDs on P0 to P3 are illuminated but by using AND # 251 the LED on P2(4) is switched off.

Stepper motor control

For the stepper motor, as with the LEDs, only the simplest of programs is provided in this text although there is plenty of scope for improved software. The following program rotates the motor in one direction continuously.

PROG. 14.6 Control of stepper motor

```
            BBC                    COMMODORE 64                  PET

10       LDA #15        10       LDA #15        10       LDA #15
20       STA 65122      20       STA 56579      20       STA 59459
30 .BEG LDA #3          30 BEG LDA #3           30 BEG LDA #3
40       STA 65120      40       STA 56577      40       STA 59471
50       JSR DEL        50       JSR DEL        50       JSR DEL
60       LDA #6         60       LDA #6         60       LDA #6
70       STA 65120      70       STA 56577      70       STA 59471
80       JSR DEL        80       JSR DEL        80       JSR DEL
90       LDA #12        90       LDA #12        90       LDA #12
100      STA 65120      100      STA 56577      100      STA 59471
110      JSR DEL        110      JSR DEL        110      JSR DEL
120      LDA #9         120      LDA #9         120      LDA #9
130      STA 65120      130      STA 56577      130      STA 59471
140      JSR DEL        140      JSR DEL        140      JSR DEL
150      ————           150      LDA 197        150      LDA 151
160      ————           160      CMP #22        160      CMP #83
170      ————           170      BEQ FIN        170      BEQ FIN
180      JMP BEG        180      JMP BEG        180      JMP BEG
190      ————           190 FIN RTS            190 FIN RTS
500 .DEL LDX #50        500 DEL LDX #50         500 DEL LDX #50
510 .L2  LDY #255       510 L2  LDY #255        510 L2  LDY #255
520 .L1  DEY            520 L1  DEY             520 L1  DEY
530      BNE L1         530      BNE L1         530      BNE L1
540      DEX            540      DEX            540      DEX
550      BNE L2         550      BNE L2         550      BNE L2
560      RTS            560      RTS            560      RTS
```

Lines 10 and 20 set four lines as output.
Lines 30 to 140 produce the output sequence required by the motor (3, 6, 12, 9).
Lines 150 to 170 are the escape routine for Commodore 64 and PET micros.
Line 180 starts the program sequence again.
Lines 500–560 provide the time delay which controls the motor speed.

Traffic light simulator

Reference to Table 12.1 provides the necessary values that must be placed in the data register to produce the most basic traffic light simulation sequence. The program below achieves this and as can be seen it is just a sequence of load and stores with a time delay between each set. It provides no new programming skill, but practice is the best of all instructors.

PROG. 14.7 Traffic light simulation

```
        BBC                    COMMODORE 64               PET

10      LDA #255       10      LDA #255        10      LDA #255
20      STA 65122      20      STA 56579       20      STA 59459
30 .BEG LDA #19        30 BEG  LDA #19         30 BEG  LDA #19
40      STA 65120      40      STA 56577       40      STA 59471
50      JSR DEL        50      JSR DEL         50      JSR DEL
60      LDA #20        60      LDA #20         60      LDA #20
70      STA 65120      70      STA 56577       70      STA 59471
80      JSR DEL        80      JSR DEL         80      JSR DEL
90      LDA #18        90      LDA #18         90      LDA #18
100     STA 65120      100     STA 56577       100     STA 59471
110     JSR DEL        110     JSR DEL         110     JSR DEL
120     LDA #49        120     LDA #49         120     LDA #49
130     STA 65120      130     STA 56577       130     STA 59471
140     JSR DEL        140     JSR DEL         140     JSR DEL
150     LDA #65        150     LDA #65         150     LDA #65
160     STA 65120      160     STA 56577       160     STA 59471
170     JSR DEL        170     JSR DEL         170     JSR DEL
180     LDA #33        180     LDA #33         180     LDA #33
185     STA 65120      185     STA 56577       185     STA 59471
190     JSR DEL        190     JSR DEL         190     JSR DEL
200     ———            200     LDA 197         200     LDA 151
210     ———            210     CMP #22         210     CMP #83
220     ———            220     BEQ FIN         220     BEQ FIN
230     JMP BEG        230     JMP BEG         230     JMP BEG
240     ———            240 FIN RTS            240 FIN  RTS
500 .DEL LDA #8        500 DEL LDA #8          500 DEL LDA #8
510     STA 31000      510     STA 50000       510     STA 800
520 .L3  LDY #255      520 L3  LDY #255        520 L3  LDY #255
530 .L2  LDX #255      530 L2  LDX #255        530 L2  LDX #255
540 .L1  DEX           540 L1  DEX             540 L1  DEX
550     BNE L 1        550     BNE L1          550     BNE L 1
560     DEY            560     DEY             560     DEY
570     BNE L2         570     BNE L2          570     BNE L2
580     DEC 31000      580     DEC 50000       580     DEC 800
590     BNE L3         590     BNE L3          590     BNE L3
600     RTS            600     RTS             600     RTS
```

Handling input signals

All the software so far in this chapter has been concerned with simply controlling external devices without regard for any incoming signals, so now let us examine how to monitor or read incoming signals and respond accordingly. Probably the most useful way of doing this is to use the circuit (Fig. 12.6) which allowed the control of a stepper motor through switches. Here, however, the software is designed to simply stop and start the motor when particular switches are opened.

PROG. 14.8 Control of stepper motor using switches

BBC			COMMODORE 64			PET		
10		LDA #15	10		LDA #15	10		LDA #15
20		STA 65122	20		STA 56579	20		STA 59459
30	BEG	LDA #128	30	BEG	LDA #128	30	BEG	LDA #128
40		AND 65120	40		AND 56577	40		AND 59471
50		BNE ROT	50		BNE ROT	50		BNE ROT
60		LDA #64	60		LDA #64	60		LDA #64
70		AND 65120	70		AND 56577	70		AND 59471
80		BNE CUT	80		BNE CUT	80		BNE CUT
90		——————	90		LDA 197	90		LDA 151
100		——————	100		CMP #	100		CMP #86
110		——————	110		BEQ FIN	110		BEQ FIN
120	CUT	LDA #0	120	CUT	LDA #0	120	CUT	LDA #0
130		STA 65120	130		STA 56577	130		STA 59471
140		JMP BEG	140		JMP BEG	140		JMP BEG
150		——————	150	FIN	RTS	150	FIN	RTS
160	ROT	LDA #3	160	ROT	LDA #3	160	ROT	LDA #3
170		STA 65120	170		STA 56577	170		STA 59471
180		JSR DEL	180		JSR DEL	180		JSR DEL
190		LDA #6	190		LDA #6	190		LDA #6
200		STA 65120	200		STA 56577	200		STA 59471
210		JSR DEL	210		JSR DEL	210		JSR DEL
220		LDA #12	220		LDA #12	220		LDA #12
230		STA 65120	230		STA 56577	230		STA 59471
240		JSR DEL	240		JSR DEL	240		JSR DEL
250		LDA #9	250		LDA #9	250		LDA #9
260		STA 65120	260		STA 56577	260		STA 59471
270		JSR DEL	270		JSR DEL	270		JSR DEL
280		LDA #64	280		LDA #64	280		LDA #64
290		AND 65120	290		AND 56577	290		AND 59471
300		BNE CUT	300		BNE CUT	300		BNE CUT
310		JMP ROT	310		JMP ROT	310		JMP ROT
500	DEL	LDX #100	500	DEL	LDX #100	500	DEL	LDX #100
510	L2	LDY #255	510	L2	LDY #255	510	L2	LDY #255
520	L1	DEY	520	L1	DEY	520	L1	DEY
530		BNE L1	530		BNE L1	530		BNE L1
540		DEX	540		DEX	540		DEX
550		BNE L2	550		BNE L2	550		BNE L2
560		RTS	560		RTS	560		RTS

Lines 10 and 20 set four lines as outputs and four as inputs.
Lines 30 to 80 test for incoming signals on P7 (128) and P6(64); P7 starts the
    motor and P6 stops the motor.

Lines 90 to 130 and 150 are the Commodore 64 and PET escape routines.

Line 140 returns the program to the start.

Lines 160 to 270 provide the sequence of pulses required to rotate the stepper motor.

Lines 280 to 300 test for a stop motor signal on P6.

Line 310 keeps the motor rotating.

Lines 500+ provide the time delay.

15

Robot arms

The word 'robot' was added to our vocabulary by a Czech playwright, Karel Capek, in a 1920 play called 'Rossums Universal Robots' or RUR for short. In the play synthetic creatures called robots strictly obeyed their masters' orders. The word 'robot' derives from a similar Czech word meaning 'forced labour'.

The term robot is one with many connotations ranging from a simple mechanical arm to the futuristic androids. Progression in this field of technology is extremely fast moving. The world total industrial robot population is increasing rapidly with Japan being the main user. There is also a fairly large population of non-industrial robots. Of the non-industrial types the robot arm is probably the most common.

The term 'robot arm' embraces a considerable diversity of devices ranging from the simple pick-and-place arms with a very limited range of movements and control, to the most modern and elaborate high technology arms with greater range of movements, greater 'intelligence' and perhaps incorporating optical and other sensors providing increased accuracy and adaptability.

Features of the robot arm

Geometry

A popular design, giving a good range of movements (i.e. degrees of freedom) and not unlike the human arm is shown in Fig. 15.1. Rotation of the waist permits a 360° access around the arm. There is movement in the shoulder, the forearm and generally two movements of the wrist— up and down and angular. Some of the cheapest robot arms are of this design but they are of Meccano-like construction. This angular design gives a fairly robust construction, but greatly complicates the software.

Figure 15.2 illustrates some of the other types of geometry.

Actuators

Various arrangements are employed for driving the arms. Pneumatic, hydraulic or electrical drive systems are employed and each offers its own advantages.

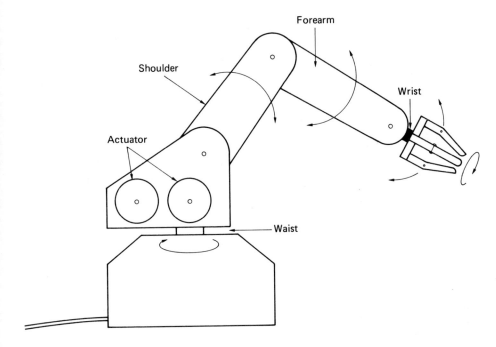

Fig. 15.1 Angular type arm

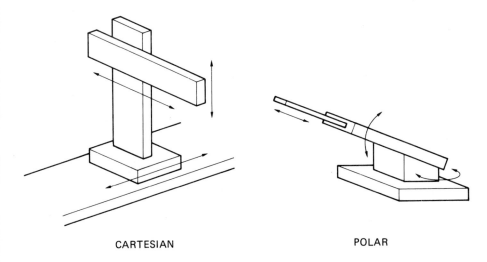

CARTESIAN   POLAR

Fig. 15.2 Arm geometries

Pneumatic – this offers the cheapest form of power and is popular with simple pick-up-and-place robots.

Hydraulic – capable of providing great force and so is a natural choice for large arms handling heavy loads.

Electrical – these drives can take various forms. (a)   Large d.c. motors are common in the heavier arms with electrical drives. They are simple to control, but some form of feedback is required for positional control. (b)   Stepper motors are a very popular means of driving small arms. They are easy to drive and offer accurate positional control without the use of feedback. A typical small arm may be driven by up to six stepper motors.

Repeatability

The ability of an arm to repeat the same movements exactly time after time is naturally an important feature of many industrial robot arms. In the cheaper, non-industrial arms this feature is not vital, but naturally desirable and frustration sets in when the arm loses positional accuracy after only four or five cycles. If you or your college or school, etc. intend purchasing an arm, examine closely the repeatability qualities, perhaps even when the arm is under some load.

Load capacity

For obvious reasons this is a crucial factor that must be considered in many industrial applications. For non-industrial applications this in many cases is not important and need not be considered.

Types of gripper

Again, as with the load capacity factor, the type of gripper or hand is very important in industrial robots, but outside these applications one tends to overlook such 'minor' details. Two common types are given in Fig. 15.3.

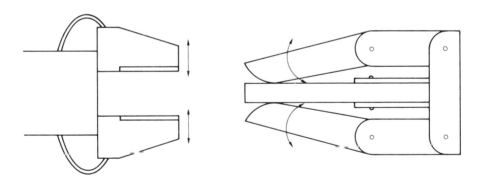

Fig. 15.3 Arm grippers

Other considerations

One consideration that is extremely important in industrial robots is the speed of operation. This is not so critical with non-industrial applications, and often tends to be ignored. A feature that is important to all robot owners is reliability, but obtaining information in this area is not easy.

Control method

The 'pick-up-and-place' type arms are probably the simplest to control and the microprocessor perhaps only controls the sequence of movements but not the magnitude of movement. Positional control with such low technology arms can be achieved by microswitches or valves that cut the power off when the arm achieves the desired position.

Arms that simulate the human arm demand positional control by the microprocessor on each axis, and this means complex software. If stepper motors are used on such arms the desired position is attained by stepping the motor through the requisite angle. Other drive systems require a feedback mechanism as a method of detecting arm position.

Of the cheaper non-industrial robot arms some have an on-board computer and others can be controlled by an external computer. Certain 'small-arms' manufacturers produce a basic arm and then supply user port adaptors and software to suit customers' microcomputers. Control simply means entering the software, connecting the robot to the user port and driving the arm is then a matter dictated by the program.

Teaching a robot

Robots arms are 'taught' a sequence of operations by either preparing a program listing of the path data or by pressing keys which control the arm directly and taking the arm through the required sequence and pressing a specified 'learn' key when each step has been completed satisfactorily.

Typical software (using the latter type of 'teaching' method) allocates keys on the microcomputer keyboard to control individual arm movements. For example, 0 and 1 could be used to open and close the gripper, W and " used to rotate the wrist, etc. Even for the novice, controlling an arm from the micro keyboard is quickly picked up. Unfortunately, the arm can easily become a plaything and nothing actually be learned about robotics.

Robot arm open loop control system— stepper motor drive

Both open and closed loop control systems are employed with the smaller robot arms. However stepper motors (providing an open loop control system) are very popular because the overall system is simpler and cheaper than closed loop systems.

A typical robot arm may have, say, six stepper motors for control of the gripper, wrist roll, wrist pitch, elbow, shoulder and the waist.

Recalling that four control lines are required to drive one stepper motor, it appears that only two motors can be controlled from the user port. Under normal conditions this is true, but electronic hardware is provided with the arm and this allows selection and control of particular motors. Figure 15.4 shows a simplified control circuit. Note that the input has three distinct parts.

  (i) motor stepping codes
 (ii) motor selection codes
(iii) a strobe bit

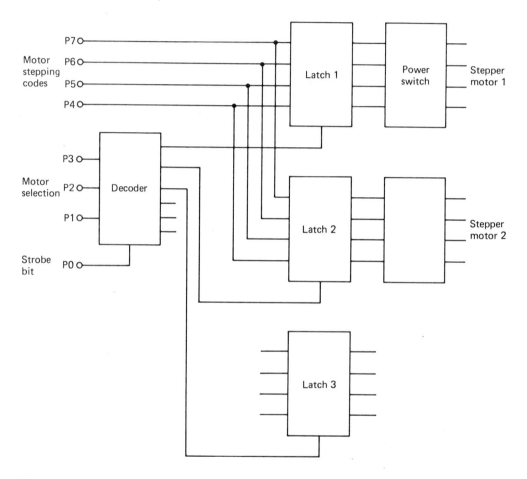

Fig. 15.4 Open loop control of stepper motors

(i) Motor stepping codes
P4 to P7 provide the standard stepper motor four bit stepping codes. These are listed below.

(ii) Motor selection codes
P1 to P3 is a three bit code used to select a particular motor. A typical set of codes might be:

Step	P7 (128)	P6 (64)	P5 (32)	P4 (16)	Denary value poked
1	0	0	1	1	48
2	0	1	1	0	96
3	1	1	0	0	192
4	1	0	0	1	144
5	0	0	1	1	48

Motor selected	P3 (8)	P2 (4)	P1 (2)	Denary value poked
1 Gripper	0	0	1	2
2 Wrist roll	0	1	0	4
3 Wrist pitch	0	1	1	6
4 Elbow	1	0	0	8
5 Shoulder	1	0	1	10
6 Waist	1	1	0	12

Note the relationship between the denary value poked to the user port and the motor selected.

(iii) The strobe bit
P0 provides a one bit strobe pulse (logic 0 or 1) which causes the electronic hardware to provide the selected motor with the stepping code on P4/P7, causing the motor to step one.

A program to control one selected motor is given below. This BASIC program can control a robot arm motor but this is far too slow and machine code is required.

```
 10 DATA192,144,48,96,0
 20 REM ROBOT
 30 INPUT"MOTOR";M
 40 POKE56579,255
 50 P=(M*2)
 60 POKE56577,P
 70 READA
 80 IF A=0 THEN 130
 90 POKE56577,PEEK(56577)ORA OR1
100 POKE56577,PEEK(56577)ANDNOT1
110 POKE56577,PEEK(56577)ANDNOTA
120 GOTO 70
130 RESTORE
140 GOTO 20
```

BBC and PET software is similar except that the data direction registers and input/output registers are located at different addresses and the BBC uses the ? operator and not peek/poke.

Line 10 provides the motor stepping codes.
      30 allows selection of a particular motor.
      40 sets all lines P0/P7 as outputs.
      50 produces the motor selection code.
      60 selects the motor.
      70 reads the stepping codes.
      80 tests if stepping codes are ended and require restoring with lines 130 and 140.
      90 outputs the stepping code and sets P0 high.
      100 outputs stepping code and clears P0.
      110 clears the stepping code data ready for the new data.

Appendix 1   Memory Maps

(a)   BBC Memory Map

Address	Description
0 to 3584	Reserved for operating system use.
3585 to 32768	User's BASIC program area Dynamic vasatle storage BASIC stack High resolution graphics
32769 to 49152	4 paged ROM's e.g. BASIC
49153 to 64512	Operating system ROM
64513 to 65280	Memory mapped input/ output
65281 to 65535	Operating system ROM

16 K RAM Model A   } 32 K RAM Model B

(c)   PET Memory Map

Address	Description
0 to 511	BASIC workign storage
512 to 633	Operating system
634 to 1023	Tape buffers
1024 to 32767	User program area
32768 to 36863	TV RAM
36864 to 49151	Expansion RAM
49152 to 59392	BASIC interpreter
59392 to 61439	Input/Output
61440 to 65535	Operating system

(b)   COMMODORE 64 MEMORY MAP

ADDRESS	DESCRIPTION
0 & 1	6510 Registers
2 to 1023	Start of memory Memory used by the operating system
1024 to 2039	Screen memory
2040 to 2047	SPRITE pointer
2048 to 40959	User's memory where BASIC or machine language programs, or both, are stored
40960 to 49151	8K CBM BASIC Interpreter
49152 to 53247	Special programs RAM area.
53248 to 53294	VIC-II
54272 to 55295	SID registers
55296 to 56296	Colour RAM
56320 to 57343	Input/Output registers (6526s)
57344 to 65535	BK CBM KERNAL Operating system

Appendix 2 ASCII codes

Decimal Code	Standard Name	Decimal Code	Standard Name	Decimal Code	Standard Name
32	SPACE	64	@	96	.
33	!	65	A	97	a
34	"	66	B	98	b
35	#	67	C	99	c
36	$	68	D	100	d
37	%	69	E	101	e
38	&	70	F	102	f
39	'	71	G	103	g
40	(	72	H	104	h
41	)	73	I	105	i
42	*	74	J	106	j
43	+	75	K	107	k
44	,	76	L	108	l
45	–	77	M	109	m
46	.	78	N	110	n
47	/	79	O	111	o
48	0	80	P	112	p
49	1	81	Q	113	q
50	2	82	R	114	r
51	3	83	S	115	s
52	4	84	T	116	t
53	5	85	U	117	u
54	6	86	V	118	v
55	7	87	W	119	w
56	8	88	X	120	x
57	9	89	Y	121	y
58	:	90	Z	122	z
59	;	91	[	123	{
60	<	92	/	124		
61	=	93	]	125	}
62	>	94	^	126	~
63	?	95	—	127	DEL

Solutions to Exercises

2.1 (a) 136
     (b) 37
     (c) 960

2.2 (a) 111011
     (b) 11010011
     (c) 100111011

2.3 (a) F3
     (b) 37
     (c) FD8

2.4 (a) 110001
     (b) 10100011
     (c) 10011101000

2.5 (a) 74
     (b) E9
     (c) 37

2.6 (a) 96
     (b) 630
     (c) 1256

PROG. 4.5

PET	COMMODORE 64	BBC

```
PET                        COMMODORE 64               BBC

10 REM BINARY COUNT        10 REM BINARY COUNT        10 REM BINARY COUNT
20 POKE 59459,255          20 POKE 56579,255          20 ?65122 = 255
30 FOR X = 0 TO 255        30 FOR X = 0 TO 255        30 FOR X = 0 TO 255
40 POKE 59471,X            40 POKE 56577,X            40 ?65120 = X
50 GOSUB 500               50 GOSUB 500               50 GOSUB 500
60 NEXT X                  60 NEXT X                  60 NEXT X
70 GOTO 30                 70 GOTO 30                 70 GOTO 30
500 REM DELAY              500 REM DELAY              500 REM DELAY
510 FOR I = 0 TO 200       510 FOR I = 0 TO 200       510 FOR I = 0 TO 200
520 NEXT I                 520 NEXT I                 520 NEXT I
530 RETURN                 530 RETURN                 530 RETURN
```

PROG. 4.6

PET	COMMODORE 64	BBC
10 REM SEQUENTIAL LEDS	10 REM SEQUENTIAL LEDS	10 REM SEQUENTIAL LED
20 POKE 59459,255	20 POKE 56579,255	20 ?65122 = 255
30 LET X = 1	30 LET X = 1	30 LET X = 1
40 POKE 59471,X	40 POKE 56577,X	40 ?65120 = X
50 GOSUB 500	50 GOSUB 500	50 GOSUB 500
60 X = X*2	60 X = X*2	60 X = X*2
70 IF X>128 THEN 30	70 IF X>128 THEN 30	70 IF X>128 THEN 30
80 POKE 59471,X	80 POKE 56577,X	80 ?65120 = X
90 GOSUB 500	90 GOSUB 500	90 GOSUB 500
100 GOTO 60	100 GOTO 60	100 GOTO 60
500 REM DELAY	500 REM DELAY	500 REM DELAY
510 FOR I = 0 TO 200	510 FOR I = 0 TO 200	510 FOR I = 0 TO 200
520 NEXT I	520 NEXT I	520 NEXT I
530 RETURN	530 RETURN	530 RETURN

PROG. 4.7

PET	COMMODORE 64	BBC
10 REM LED PAIRS	10 REM LED PAIRS	10 REM LED PAIRS
20 POKE 59459,255	20 POKE 56579,255	20 ?65122 = 255
30 POKE 59471,3	30 POKE 56577,3	30 ?65120 = 3
40 GOSUB 500	40 GOSUB 500	40 GOSUB 500
50 POKE 59471,12	50 POKE 56577,12	50 ?65120 = 12
60 GOSUB 500	60 GOSUB 500	60 GOSUB 500
70 POKE 59471,48	70 POKE 56577,48	70 ?65120 = 48
80 GOSUB 500	80 GOSUB 500	80 GOSUB 500
90 POKE 59471,192	90 POKE 56577,192	90 ?65120 = 192
100 GOSUB 500	100 GOSUB 500	100 GOSUB 500
110 GOTO 20	110 GOTO 20	110 GOTO 20
500 REM DELAY	500 REM DELAY	500 REM DELAY
510 FOR I = 0 TO 200	510 FOR I = 0 TO 200	510 FOR I = 0 TO 200
520 NEXT I	520 NEXT I	520 NEXT I
530 RETURN	530 RETURN	530 RETURN

PROG. 6.4

BBC

```
10 REM STEPPER MOTOR CONTROL
20 REM SET OUTPUT LINES
30 ?65122=255
40 INPUT "NUMBER OF REVS",R
50 INPUT "DELAY",D
60 INPUT "DIRECTION OF ROTATION",D$
70 IF D$="C" THEN 200
80 IF D$="A" THEN 400
200 N=0
210 ?65120=3
220 GOSUB 600
230 ?65120=6
240 GOSUB 600
250 ?65120=12
260 GOSUB 600
270 ?65120=9
280 GOSUB 600
290 N=N+1
300 S=(N/12)
310 IF S=R THEN 330
320 GOTO 210
330 END
400 N=0
410 ?65120=9
420 GOSUB 600
430 ?65120=12
440 GOSUB 600
450 ?65120=6
460 GOSUB 600
470 ?65120=3
480 GOSUB 600
490 N=N+1
500 S=(N/12)
510 IF S=R THEN 330
520 GOTO 410

600 REM DELAY
610 FOR I=0 TO D
620 NEXT I
630 RETURN
```

COMMODORE 64

```
10 REM STEPPER MOTOR CONTROL
20 REM SET OUTPUT LINES
30 POKE56579,255
40 INPUT "NUMBER OF REVS",R
50 INPUT "DELAY",D
60 INPUT "DIRECTION OF ROTATION",D$
70 IF D$="C" THEN 200
80 IF D$="A" THEN 400
200 N=0
210 POKE56577,3
220 GOSUB 600
230 POKE56577,6
240 GOSUB 600
250 POKE56577,12
260 GOSUB 600
270 POKE56577,9
280 GOSUB 600
290 N=N+1
300 S=(N/12)
310 IF S=R THEN 330
320 GOTO 210
330 END
400 N=0
410 POKE56579,9
420 GOSUB 600
430 POKE56577,12
440 GOSUB 600
450 POKE56577,6
460 GOSUB 600
470 POKE56577,3
480 GOSUB 600
490 N=N+1
500 S=(N/12)
510 IF S=R THEN 330
520 GOTO 410

600 REM DELAY
610 FOR I=0 TO D
620 NEXT I
630 RETURN
```

PET

```
10 REM STEPPER MOTOR CONTROL
20 REM SET PORT
30 POKE59459,255
40 INPUT "NUMBER OF REVOLUTIONS";NR
50 INPUT "DELAY";D
60 INPUT "DIRECTION OF ROTATION";D$
70 IF D$="C" THEN S=1:E=4:F=1
80 IF D$="A" THEN S=4:E=1:F=-1
90 :
100 REM TWELVE PULSE SEQUENCES/REV
110 NT=NR*12
120 FOR I=1 TO NT
130 FOR J=S TO E STEP F
140 GOSUB 600
150 N(1)=9:N(2)=12:N(3)=6:N(4)=3
160 GOSUB 600
170 NEXT J
180 NEXT I
190 :
200 INPUT "ANOTHER GO Y/N";G$
210 IF G$="Y" THEN 40
220 STOP
230 :
500 REM....DELAY....
510 FOR T=1 TO D
520 NEXT T
530 RETURN
540 :
600 REM...OUTPUT VALUE...
610 POKE59471,N(J)
620 RETURN
```

The BBC and Commodore 64 programs are simpler but less efficient.

PROG. 6.5

The solution is as for program 6.2 except that the sequence of numbers required at the user port is now 48, 96, 192, 144.

PROG. 7.4

PET

```
 10 INPUT"REVS";R
 20 POKE59459,127
 30 POKE59471,1
 40 N=0
 50 A=TI
 60 IF PEEK(59471)=1 THEN 60
 70 IF PEEK(59471)=129 THEN 70
 80 N=N+1
 90 PRINT N;"REVS"
100 IF N=R THEN 120
110 GOTO 60
120 B=TI
130 POKE59471,0
140 C=(B-A)/3600
150 S=1NT(R/C)
160 PRINT S;"REVS/MIN"
170 END
```

COMMODORE 64

```
 10 INPUT"REVS";R
 20 POKE56579,127
 30 POKE56577,1
 40 N=0
 50 A=TI
 60 IF PEEK(56577)=1 THEN 60
 70 IF PEEK(56577)=129 THEN 70
 80 N=N+1
 90 PRINT N;"REVS"
100 IF N=R THEN 120
110 GOTO 60
120 B=TI
130 POKE56577,0
140 C=(B-A)/3600
150 S=INT(R/C)
160 PRINT S;"REVS/MIN"
170 END
```

BBC

```
 10 INPUT"REVS";R
 20 ?65122=127
 30 ?65120=1
 40 N=0
 50 A=TIME
 60 IF ?65120=1 THEN 60
 70 IF ?65120=129 THEN 70
 80 N=N+1
 90 PRINT N;"REVS"
100 IF N=R THEN 120
110 GOTO 60
120 B=TIME
130 ?65120=0
140 C=(B-A)/6000
150 S=INT(R/C)
160 PRINT S;"REVS/MIN"
170 END
```

Note – Line 140 is the same for the Commodore 64 and PET the time is given in 'Jiffies' (1/60th s) whereas on the BBCB it is given in 1/100th s.

PROG. 8.3

PET

```
10 REM SQUAREWAVE
20 POKE59459,255
30 POKE59471,255
40 FOR I=0 TO 500
50 NEXT I
60 POKE59471,0
70 FOR I=0 TO 500
80 NEXT I
90 GOTO 30
```

COMMODORE 64

```
10 REM SQUAREWAVE
20 POKE56579,255
30 POKE56577,255
40 FOR I=0 TO 500
50 NEXT I
60 POKE56577,0
70 FOR I=0 TO 500
80 NEXT I
90 GOTO 30
```

BBC

```
10 REM SQUAREWAVE
20 ?65122=255
30 ?65120=255
40 FOR I=0 TO 500
50 NEXT I
60 ?65120=0
70 FOR I=0 TO 500
80 NEXT I
90 GOTO 30
```

As the program stands it is ideal for an XY plotter, but removal of lines 40, 50, 70 and 80 produces a very good trace on an oscilloscope.

PROG. 8.4

PET

```
10  REM SINEWAVE
20  POKE59459,255
30  FOR X=O TO 255
40  Y=128+INT(128*SIN(X*π/128)
50  POKE59471,Y
60  NEXT X
70  GOTO 30
```

COMMODORE 64

```
10  REM SINEWAVE
20  POKE56579,255
30  FOR X=O TO 255
40  Y=128+INT(128*SIN(X*π/128)
50  POKE56577,Y
60  NEXT X
70  GOTO 30
```

BBC

```
10  REM SINEWAVE
20  ?65122=255
30  FOR X=O TO 255
40  Y=128+INT(128*SIN(X*π/128)
50  ?65120=Y
60  NEXT X
70  GOTO 30
```

Assuming the DAC can only produce positive voltages, and because the sinewave moves positively then negatively over one cycle it is necessary to shift the sinewave vertically so that its centre is very positive as shown in the diagram. This is done by adding 128 in line 40 of the program.

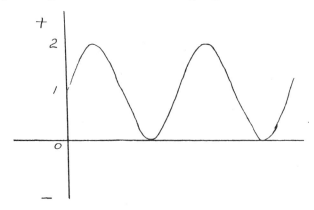

9.1   (a) 0101
         (b) 01000010
         (c) 00010000
         (d) 00000010

9.2   (a) 1111
         (b) 11110011
         (c) 11110111
         (d) 01111110

Index